Utilizing Innovative Technologies to Address the Public Health Impact of Climate Change:

Emerging Research and Opportunities

Debra Weiss-Randall
Florida Atlantic University, USA

A volume in the Advances in Environmental Engineering and Green Technologies (AEEGT) Book Series

IGI Global
DISSEMINATOR OF KNOWLEDGE

Published in the United States of America by
IGI Global
Engineering Science Reference (an imprint of IGI Global)
701 E. Chocolate Avenue
Hershey PA, USA 17033
Tel: 717-533-8845
Fax: 717-533-8661
E-mail: cust@igi-global.com
Web site: http://www.igi-global.com

Library of Congress Cataloging-in-Publication Data

Names: Weiss-Randall, Debra, 1954- author.
Title: Utilizing innovative technologies to address the public health impact of climate change : Emerging research and opportunities / by Debra Weiss-Randall.
Description: Hershey, PA : Engineering Science Reference, [2018]
Identifiers: LCCN 2017017589| ISBN 9781522534143 (hardcover) | ISBN 9781522534150 (ebook)
Subjects: LCSH: Climatic changes--Health aspects--United States. | Global warming--Health aspects--United States. | Public health--Technological innovation--United States. | Environmental health--Technological innovations--United States.
Classification: LCC RA804 .W45 2018 | DDC 333.71/4--dc23 LC record available at https://lccn.loc.gov/2017017589

This book is published in the IGI Global book series Advances in Environmental Engineering and Green Technologies (AEEGT) (ISSN: 2326-9162; eISSN: 2326-9170)

British Cataloguing in Publication Data
A Cataloguing in Publication record for this book is available from the British Library.

Advances in Environmental Engineering and Green Technologies (AEEGT) Book Series

ISSN:2326-9162
EISSN:2326-9170

MISSION

Growing awareness and an increased focus on environmental issues such as climate change, energy use, and loss of non-renewable resources have brought about a greater need for research that provides potential solutions to these problems. Research in environmental science and engineering continues to play a vital role in uncovering new opportunities for a "green" future.

The **Advances in Environmental Engineering and Green Technologies (AEEGT)** book series is a mouthpiece for research in all aspects of environmental science, earth science, and green initiatives. This series supports the ongoing research in this field through publishing books that discuss topics within environmental engineering or that deal with the interdisciplinary field of green technologies.

COVERAGE

- Policies Involving Green Technologies and Environmental Engineering
- Air Quality
- Renewable energy
- Contaminated Site Remediation
- Cleantech
- Alternative Power Sources
- Sustainable Communities
- Waste Management
- Green Technology
- Pollution Management

IGI Global is currently accepting manuscripts for publication within this series. To submit a proposal for a volume in this series, please contact our Acquisition Editors at Acquisitions@igi-global.com or visit: http://www.igi-global.com/publish/.

The Advances in Environmental Engineering and Green Technologies (AEEGT) Book Series (ISSN 2326-9162) is published by IGI Global, 701 E. Chocolate Avenue, Hershey, PA 17033-1240, USA, www.igi-global.com. This series is composed of titles available for purchase individually; each title is edited to be contextually exclusive from any other title within the series. For pricing and ordering information please visit http://www.igi-global.com/book-series/advances-environmental-engineering-green-technologies/73679. Postmaster: Send all address changes to above address.

Titles in this Series

For a list of additional titles in this series, please visit:
https://www.igi-global.com/book-series/advances-environmental-engineering-green-technologies/73679

Advanced Nanomaterials for Water Engineering, Treatment, and Hydraulics
Tawfik A. Saleh (King Fahd University of Petroleum and Minerals, Saudi Arabia)
Engineering Science Reference • ©2017 • 384pp • H/C (ISBN: 9781522521365) • US $200.00

Handbook of Research on Inventive Bioremediation Techniques
Jatindra Nath Bhakta (University of Kalyani, India)
Engineering Science Reference • ©2017 • 624pp • H/C (ISBN: 9781522523253) • US $260.00

Handbook of Research on Entrepreneurial Development and Innovation Within Smart Cities
Luisa Cagica Carvalho (Universidade Aberta, Portugal & CEFAGE - Universidade de Évora, Portugal)
Information Science Reference • ©2017 • 661pp • H/C (ISBN: 9781522519782) • US $235.00

Applied Environmental Materials Science for Sustainability
Takaomi Kobayashi (Nagaoka University of Technology, Japan)
Information Science Reference • ©2017 • 416pp • H/C (ISBN: 9781522519713) • US $205.00

Environmental Issues Surrounding Human Overpopulation
Rajeev Pratap Singh (Banaras Hindu University, India) Anita Singh (University of Allahabad, India) and Vaibhav Srivastava (Banaras Hindu University, India)
Information Science Reference • ©2017 • 325pp • H/C (ISBN: 9781522516835) • US $200.00

Reconsidering the Impact of Climate Change on Global Water Supply, Use, and Management
Prakash Rao (Symbiosis International University, India) and Yogesh Patil (Symbiosis International University, India)
Information Science Reference • ©2017 • 430pp • H/C (ISBN: 9781522510468) • US $215.00

For an enitre list of titles in this series, please visit:
https://www.igi-global.com/book-series/advances-environmental-engineering-green-technologies/73679

IGI Global
DISSEMINATOR OF KNOWLEDGE

701 East Chocolate Avenue, Hershey, PA 17033, USA
Tel: 717-533-8845 x100 • Fax: 717-533-8661
E-Mail: cust@igi-global.com • www.igi-global.com

Table of Contents

Foreword

Climate change will be one of the greatest issues for public health this century. In fact, climate change is a broad category of many distinct issues. Terrestrial aspects range from direct effects of record high temperatures, to changes to the ecosystem, new diseases and pathways. In the ocean, beyond the warmer temperatures and changing currents, the change in acidity level, measured as pH, is already showing tremendous impact. As the alkalinity level moves lower towards the acidic, shellfish and corals will by physically threatened, as their structures will be unable to form due to fundamental chemistry.

As the glaciers and ice sheets on land in the polar regions melt, sea level will rise globally affecting coastal communities from Miami to the Maldives and from Boston to Bangladesh. The rising sea will have its greatest impact not in the ocean, but rather, where we live, on land, adding to the range of issues affecting people of diverse economic, social, political, and ethnic backgrounds.

In this ambitious work, Dr. Debra Weiss-Randall has taken on this extremely important area, covering some of the most prominent issues where climate change will impact human health. Dr. Weiss-Randall has done an admirable job of researching, compiling, and analyzing a vast body of knowledge on the subject. She suggests ways that innovative technologies may enable better public health.

One of her focus areas is vulnerable populations, the challenge of caring for those most at risk from the impacts; Dr. Weiss-Randall describes initiatives that hopefully will achieve environmental justice for these populations, to empower them to develop resilience and to insure that they have protection from climate change impacts. Communities will change greatly in the coming decades. Impacts will not respect national borders or even boundaries of localized governance.

As the earth warms – an unstoppable occurrence in the near to medium term-- these will be powerful challenges. Looking ahead over the coming decade or two, societies will have to begin thinking creatively.

This book should be a welcome addition both in academia and in the professional arena of public health, both for the practitioner as well as administrators.

Adapting to the challenges of climate change is now squarely in our hands. It is impossible to fully predict what the future will be like. We have finished the ten thousand years of relative stability – geologically known as the *Holocene* – and are now in the *Anthropocene*, the era characterized by the impact of our species.

Technology is the dominant aspect of our effect on the planet aside from our sheer numbers, now some seven billion, projected to be almost ten billion by mid-century. It seems fitting and inevitable that we will have to look to technology to help humanity cope with this changing climate. This book should be a useful textbook or guide for those in both public health and environmental science.

John Englander
Englander & Associates, USA

John Englander *is an oceanographer and the author of High Tide on Main Street: Rising Sea Level and the Coming Coastal Crisis (The Science Bookshelf, 2nd Edition 2013) and the weekly newsletter www.sealevelrisenow.com. John also has his own website, at www.johnenglander.net.*

Preface

OVERVIEW

Climate change is the most urgent issue facing the world today. It has been described as "the most fateful battle in human history" (McKibben, 2013). World leaders have been urging people "to wake up to the fierce urgency of now. It's our responsibility to ensure a livable planet—for us and for all future generations" (Kim, 2016). Prominent scientists are predicting that, if current energy trends continue unabated, the planet will be unfit for human habitation in 100 years (Hawking, 2017; Stockton, 2016).

There is overwhelming scientific evidence that global warming is due to human-made greenhouse gas emissions, caused by the burning of fossil fuels. Most of the recorded warming of the Earth's surface temperatures has occurred in the last 35 years, with 16 of the 17 hottest years reported in the period 2001-2016; the hottest year on record was 2016 (NASA, 2017; NOAA, 2017). In addition, in 2016 the average Arctic sea ice extent was the smallest since record-keeping began (Richter-Menge et al., 2016). In the first quarter of 2017, the carbon dioxide (CO_2) concentration in the atmosphere soared to more than 400 ppm (parts per million). Sea levels are projected to rise one to two meters (roughly 3 to 7 feet) by 2100. It is vitally important that the nations of the world reduce CO_2 emissions drastically to slow down global warming.

In December of 2015 an historic agreement was signed in Paris, committing the nations of the world to reducing fossil fuel emissions. Europe has taken the lead in switching over to renewable clean energy sources such as solar, wind, water, and thermal. However, much more needs to be done! The atmospheric concentration of CO_2 was at a dangerously high level in the first quarter of 2017. In the words of Dr. James Hansen (2012):

Adding CO_2 [carbon dioxide] to the air is like throwing another blanket on the bed. It reduces earth's heat radiation to space so there is a temporary energy imbalance—more energy is going in than going out. The earth's energy imbalance is...equivalent to exploding 400,000 Hiroshima atomic bombs per day 365 days per year. That's how much extra energy earth is gaining each day....we must reduce CO_2 to 350ppm... to restore energy balance and prevent further warming.

The public health impacts of global warming affect every aspect of life on the planet. It could well be argued that what has already occurred is "just the tip of the iceberg." Recent heat waves have seen temperatures of 54 degrees Celsius (129.3 degrees Fahrenheit). Severe air pollution has endangered lung health, while water contamination has led to a resurgence of cholera. Droughts and desertification have brought about food scarcity and malnutrition, as well as displacement of climate refugees. Vector-borne infections such as Zika and Lyme Disease have soared. More frequent and severe natural disasters have caused huge casualties and economic losses.

Today new technologies can be used to mitigate the adverse health consequences of climate change. Emergency preparedness is one area that has benefited from such innovations. Examples include: 1) global monitoring systems that help to warn populations of impending disasters; 2) mapping technologies that identify vulnerable populations so that they can be transported to safety during emergencies; and 3) drones equipped with the latest Geographic Information Systems (GIS) that help to safely coordinate rescue operations during natural disasters and other emergencies.

Sustainable development is another area that benefits from technology. In urban areas, "green teams" are developing environmentally friendly building projects, and retrofitting existing buildings with energy-efficient lighting, heating, air-conditioning and ventilation (HVAC) systems. In agriculture, sustainable practices such as no-till farming and drip irrigation are being implemented to reduce water usage and GHG emissions.

WHERE THIS TOPIC FITS

This book fills a gap in the literature on global warming; it provides a health education perspective. In a previous book (Weiss, 2016), I discussed how Internet-based health education is being used to educate the public about disease prevention and management, and to train health workers to fill critical shortages in developing nations. This book addresses how health education, used in conjunction with technology, can reduce casualties during emergencies and mitigate the adverse health consequences of climate change. It also may be useful in helping planners develop health education sections of Climate Change Action Plans.

TARGET AUDIENCE

This book's particular scope and perspective will benefit 1) university and college faculty, researchers, undergraduate and graduate students, adult vocational students, librarians, and academic leaders; 2) health educators; 3) urban planners and sustainability coordinators; 4) managers and employees at NGOs involved in sustainable development; and 5) IT developers working with the aforementioned four categories.

ORGANIZATION OF THE BOOK

This book is divided into eight chapters, each of which provides a reference list. Chapter 1 presents an overview of global warming, a discussion of key climate change indicators, and a section on global trends and impacts. Chapter 2 explores the major health impacts of climate change—temperature-related illness and death, air pollution (indoor and outdoor), food safety, nutrition, and availability, water-borne diseases and sanitation, vector-borne diseases, natural hazards, and mental health and well-being. Chapter 3 looks at the impact on vulnerable populations, and how technologies can be used to help protect them from the adverse effects of climate change. Chapter 4 explores the history and principles of environmental justice, whereby all peoples receive equal protection from environmental consequences of climate change. Case examples are included. Chapter 5 describes the Diffusion of Innovations (DOI) Theoretical Framework, a widely used framework of social change that is the model of choice for convincing people to adopt innovations that can help to fight climate change; Germany is using DOI to promote the switch from traditional vehicles to electric vehicles (EVs). Chapter 6 focuses on sustainable development, based on the 17 Sustainable Development Goals (SDGs) for the world adopted by the United Nations in 2015. Chapter 7 discusses strategies for cultivating climate resilience; it includes case examples. Chapter 8 discusses climate change solutions for adapting to and mitigating the impacts of climate change.

CONCLUSION

The vast majority of scientists around the world agree that global warming is occurring as a result of human-induced fossil fuel emissions released into the atmosphere. Emissions of CO_2, methane, nitrous oxide, and fluorinated gases trap heat in the atmosphere and create "the greenhouse effect." As a result of the greenhouse effect, global mean temperatures have been rising steadily, with record high temperatures in the past three years; 2016 was the hottest year on record since 1880.

The public health threats unleashed by global warming include upsurges in food poisoning, infectious diseases, undernourishment, and water contamination, with flooding, droughts, extreme heat waves, and more frequent natural disasters causing casualties and property damage. King tides have caused street flooding in coastal areas, resulting in saltwater intrusion in the water table. The sea level is rising at an alarming rate; conservative projections put it at a meter (39.37 inches) by 2100, but it could well be greater than that, endangering coastal cities around the world. Arctic air temperatures have been increasing at *double* the global rate (Richter-Menge et al., 2016), causing Arctic sea ice to melt. The bright white Arctic sea ice serves as a giant reflector, sending solar energy back into space; as the ice melts, it becomes seawater, which absorbs and retains heat. Furthermore, the great land ice masses on Greenland and Antarctica are melting at an alarming rate; if something is not done to mitigate it, there will be catastrophic inundation of the earth!

Technological innovations play a key role in the battle against climate change. Carbon pricing, carbon taxes, carbon capture and geologic sequestration are all tools to help us to become carbon-neutral (having a net-zero carbon footprint). Switching to renewable clean energy sources requires a change in mindset, which can be facilitated through the social change framework, Diffusion of Innovations (Rogers, 2003).

This book provides guideposts for county and municipal managers of natural resources, sustainability coordinators, chief resilience officers, and certified health education specialists (CHES) to take charge of their community's adaptation to climate change, while also seeking long-term solutions to mitigate climate change. It is a welcome addition to libraries for university and college faculty, students, and the student health center. Managers at NGOs also may find this book to be a useful addition to their library.

Climate change has created a new normal, and it is now up to us to adapt to it so that our children and grandchildren will be able to enjoy this beautiful earth. I hope that the book's health education perspective on climate change will be helpful to all those studying global warming, or working on solutions to it, in their communities here in the United States and abroad.

Debra Weiss-Randall
Florida Atlantic University, USA

REFERENCES

Hansen, J. (2012). *Why I Must Speak Out about Climate Change* [TED Talk]. Available at: http://csas.ei.columbia.edu/in-the-news/videos/

Hawking, S. (2017). *Stephen Hawking: Expedition New Earth* [Film]. Academic Press.

Kim, J. Y. (2016). *Remarks by World Bank Group President Jim Yong Kim at Climate Action Summit 2016*. Washington, DC: Climate Action Summit. Available at: http://www.worldbank.org/en/news/speech/2016/05/05/remarks-world-bank-group-president-jim-yong-kim-climate-action-summit

McKibben, B. (2013). *Do the Math (the movie)*. New York: 350.org & PF Pictures.

NASA. (2017). *NASA, NOAA data show 2016 hottest year on record*. Available at: https://www.nasa.gov/press-release/nasa-noaa-data-show-2016-warmest-year-on-record-globally

National Oceanic and Atmospheric Administration (NOAA). (2017). *2016 marks the third consecutive year of record warmth for the globe*. Washington, DC: The Department of Commerce. Available at: http://www.noaa.gov/stories/2016-marks-three-consecutive-years-of-record-warmth-for-globe

Richter-Menge, J., Overland, J. E., & Mathis, J. (2016). *Arctic report card of the National Oceanographic and Atmospheric Administration (NOAA): Executive summary*. Washington, DC: U.S. Department of Commerce. Available at: http://www.arctic.noaa.gov/Report-Card/Report-Card-2016/ArtMID/5022/ArticleID/270/Executive-Summary

Rogers, E. M. (2003). *Diffusion of Innovations* (5th ed.). New York: Free Press.

Stockton, N. (2016, September). Elon Musk announces his plan to colonize Mars and save humanity. *Science*, 27.

Weiss, D. (2016). Assessing Online Courses in Health Education: Training a 21st Century Health Workforce. In V. Wang (Ed.), Handbook of Advancing Health Education through Technology. IGI Global.

Acknowledgment

I would like to thank Lindsey Wertman, Courtney Tychinsky, Jan Travers and Jordan Tepper, my editors at IGI Global, for their assistance and patience during my writing of this book. I am grateful to Melissa Wagner in marketing, for her help with the book cover. I would also like to thank my friend John Englander, who wrote the Foreword to this book. John is an oceanographer who has travelled the globe documenting the effects of global warming on sea ice and land ice. He has given me a deeper understanding of the sea level rise issue.

I am grateful for my experience last year as a U.S. EPA Project Manager on an environmental justice grant, under the umbrella of the Unitarian Universalist Fellowship of Boca Raton (UUFBR). It gave an added dimension to my work on health equity. We need more NGOs committed to environmental justice like UUFBR and its dedicated leader, Reverend Harris Riordan. I hope this book inspires readers to renew their dedication to fighting the battle against climate change. Many thanks to all those who have paved the way for us!

Debra Weiss-Randall
Florida Atlantic University, USA

Chapter 1
Climate Change Overview

ABSTRACT

There is overwhelming scientific evidence that we are experiencing global warming, and that is it due to human-made greenhouse gas emissions, not to a "natural" cycle. Two key indicators of climate change had record-breaking years in 2016: the global mean surface temperature was the highest since recording began in 1880, and the average Arctic sea ice extent was the smallest annual average since record-keeping began in 1979. The greenhouse effect, caused by the burning of fossil fuels, has accelerated as carbon dioxide concentration in the atmosphere has soared to more than 400 parts per million (ppm). As a result of global warming, sea levels are projected to rise at least one-meter (39.4 inches), possibly two meters (78.7 inches), by 2100. It is vitally important that the nations of the world reduce CO_2 emissions to slow down global warming. This chapter gives an overview of domestic and global trends in, and impacts from, climate change.

INTRODUCTION

"Yes, Virginia, there *is* climate change." Despite the protestations of some politicians, there is overwhelming scientific evidence that we are experiencing global warming, and that is it due to human-made carbon emissions, not to a "natural" cycle (NOAA, 2017, NASA, 2017, USGCRP, 2016, Union of Concerned Scientists, 2017). Rather than burying our heads in the sand, we need to stand up and work together to adapt to climate change, mitigate its future impacts, and cultivate resilience. The long-term goal is to create a

DOI: 10.4018/978-1-5225-3414-3.ch001

sustainable world, one that uses renewable energy sources rather than fossil fuels, that has enough clean water and food for all, and where everyone has access to decent sanitation and living conditions. As a first step in planning for a sustainable future, we need to understand the breadth and scope of the threats facing our planet from climate change. The global challenges of rising temperatures, rising sea levels, and increased frequency of extreme weather events are creating economic, geographic, and health impacts felt around the world.

The use of technical innovations in many fields-- environmental science, healthcare, architecture, agriculture, and hydrology, to name a few—will make the job of creating a sustainable world significantly easier to achieve. In each field, when a technology is invented, the target users have to be educated about the innovation and agree to adopt it. The new technology has to fit in with the knowledge, skills, and culture of those employing it. The adoption of new technologies can be accomplished using the Diffusion of Innovations (DOI) Theory, which began as a tool to encourage farmers to adopt more efficient technologies; it has evolved into one of the most widespread theoretical frameworks for social change. This book uses the DOI framework as an underpinning for addressing the health impacts of climate change, and arriving at a new world order that uses advanced technology to achieve sustainability.

This first chapter presents an overview of global warming and greenhouse gas emissions, key indicators of climate change, regional trends in the United States, and global trends and impacts. Chapters 2 and 3 explore the major health impacts of climate change, and their impact on vulnerable populations. Chapter 4 explores climate change and environmental justice, while Chapter 5 describes the DOI Framework. Chapter 6 focuses on sustainable development, based on the 17 Sustainable Development Goals (SDGs) for the world adopted by the United Nations in 2015, and Chapter 7 discusses strategies for cultivating resilience. Chapter 8 discusses climate change solutions for adapting to and mitigating the impacts of climate change.

GLOBAL WARMING AND GREENHOUSE GAS EMISSIONS

The Earth's average temperature has been rising over the past century; this phenomenon has been confirmed by many recorded observations of air and water temperatures, sea level, and ice. According to the Intergovernmental Panel on Climate Change (IPCC), the leading international body for the assessment of climate change. the rate of change is unprecedented in human history and "human influence has been the dominant cause of this observed warming" (IPCC, 2014). The IPCC was established by the United Nations Environment Programme (UNEP) in 1988; every few years, it convenes climate experts from around the world to develop comprehensive assessments based on the latest peer-reviewed scientific/technical literature.

In the United States, the government has acknowledged that climate change poses "a significant health threat" to Americans (USGCRP, 2016). Climate change has caused changes in the weather, including rising temperatures and more frequent extreme weather events, but the two terms should not be confused: *weather* is the state of the atmosphere at any given time and place. *Climate* is the average weather conditions over many decades. The weather may change in minutes or hours, but a change in climate requires multiple decades to centuries or longer.

Global warming is a result of "the greenhouse effect," whereby greenhouse gases (GHG)—carbon dioxide, methane, nitrous oxide, and fluorinated gases—trap heat in the atmosphere. In 1896, the greenhouse effect was described by Swedish scientist Svante Arrhenius, who observed that fossil fuel combustion may result in global warming (Maslin, 2004); Arrhenius cited earlier heat studies by the French mathematician Joseph Fourier (Fourier, 1824). In 1917, the inventor Alexander Graham Bell stated that the unchecked burning of fossil fuels would lead to a "sort of greenhouse effect" and global warming. A man way ahead of his time, Bell already observed the depletion of natural resources 100 years ago (Surtees, 2003). The warming effect of GHG has kept the average surface temperature of the planet at 15 degrees Celsius (59 degrees Fahrenheit); without this greenhouse effect, the average surface temperature would be -18 degrees Celsius (0 degrees Fahrenheit) (Lang, 2010). However, the industrial age has accelerated the release of GHG to the point that the world is warming at an alarming pace. Rising GHG concentrations have contributed to rising temperatures, changes in precipitation, more frequent extreme weather events, and rising sea levels.

The major GHG is carbon dioxide (CO_2), which accounted for 81 percent of all GHG emissions in the United States in 2014 (U.S. EPA, 2016). Globally, CO_2 accounted for 64 percent of all human-induced global warming; its concentration in the atmosphere is 40 percent higher than it was in the pre-industrialization world (European Commission, 2017). CO_2 concentrations in the atmosphere are monitored closely, as they are regarded as the driver behind rising temperatures. Figure 1 shows the concentration of atmospheric CO_2 at the observatory in Mauna Loa, Hawaii.

Figure 1. Atmospheric CO_2 concentrations 1960-March 2017
Source: Scripps Institution of Oceanography (2017).

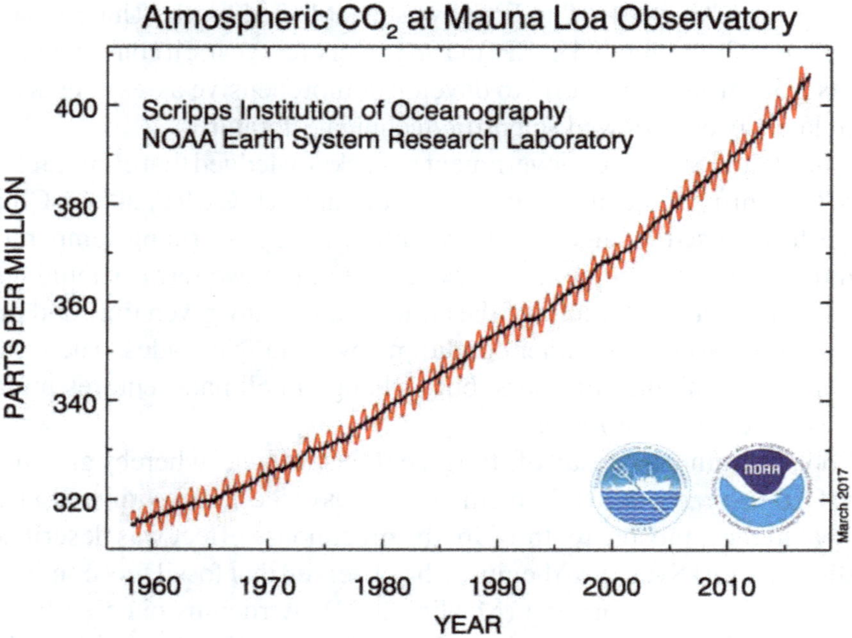

The concentration of CO_2 in the atmospheric topped 400 parts per million (ppm) for the first time in the modern record on May 9, 2013. The atmospheric CO_2 concentration was projected to continue to increase 2 ppm per year unless measures were taken to stop it (McKibben, 2013). At this writing (March 2017), the latest Mauna Loa reading posted by NOAA showed a mean monthly CO_2 concentration of 406.42 for February 2017 (Scripps Institution of Oceanography, 2017).

Carbon dioxide is released into the atmosphere from the burning of fossil fuels (coal, natural gas, and oil), solid waste, trees and wood products, and also as a result of the manufacture of cement and other chemical reactions (U.S. EPA, 2016). Deforestation in Brazil, which has eliminated almost 20 percent of the Amazon rainforest in the past 40 years, is a major contributor to CO_2 emissions, as the rainforest sequesters carbon, which is released when the trees are burned (Wallace, 2017). In addition, the burning of biomass fuel (charcoal and wood) in sub-Saharan Africa account for more than a fifth of the global CO_2 emissions (Ichoku et al., 2016).

Methane (CH_4) accounted for 11 percent of all GHG emissions in the United States in 2014, and 17 percent of human-induced global warming (European Commission, 2017). In its pure form, methane is more than 200 times more effective at trapping heat than is carbon dioxide (Englander, 2013/14). Methane is emitted during the production and transport of coal, natural gas, and oil, as well as during livestock and other agricultural practices and by the decay of organic waste in municipal solid waste landfills (U.S. EPA, 2016). There also is an enormous reservoir of undersea methane that has the potential to cause devastation to the planet. Fifty-five million years ago, a gigantic release of undersea methane caused global warming that raised temperatures 11 degrees Celsius over 20,000 years, resulting in sea levels 400 feet higher than they are today. Today the biggest source of methane is underground in the Arctic permafrost and in the seabed. As the plant warms, permafrost melts, releasing the methane. Oceanographer John Englander reports that there are lakes in Europe and Asia that have so much released methane under the ice, when people go to drill holes for ice fishing, they can ignite them. Methane clathrates form in the slushy ice on the sea floor when methane escapes from the earth and hits the cold deep seawater. It is estimated that there is more methane in this form on the seabeds than there is petroleum in the world. As the oceans warm, this methane will be released as bubbles, greatly accelerating the greenhouse effect (Englander, 2013/14).

Nitrous oxide (N_2O), which is emitted during agricultural and industrial activities, as well as during combustion of fossil fuels and solid waste, contributed 6 percent of the GHG emissions in the United States in 2014 (U.S. EPA, 2016), as well as 6 percent of human-induced global warming (European Commission, 2017). Fluorinated gases (F-gases)--hydrofluorocarbons used in refrigerants and air conditioning, perfluorocarbons used in electronics, cosmetics, and pharmaceuticals, sulfur hexafluoride used for insulation, and nitrogen trifluoride used in microchips--are synthetic, powerful greenhouse gases emitted from industry that contributed 3 percent of GHG emissions in 2014 (U.S. EPA, 2016). Although F-gases are a small percentage of total GHG emissions, they can remain in the atmosphere for thousands of years; concern over their impact has led the EU to pass a regulation to cut the EU's F-gas emissions by two-thirds in 2030 compared with 2014 levels (European Commission, 2017).

Globally, the top emitter of CO_2 is China (28 percent), followed by the United States (14.3 percent), India (6.1 percent), Russian Federation (4.7 percent), Japan (3.3 percent), Germany (2.0 percent), the Islamic Republic of Iran (1.8 percent), Saudi Arabia (1.6 percent), the Republic of Korea (1.6

percent), and Canada (1.5 percent). China and the U.S. together account for more than one-third of the world's emissions (Boden & Andres, 2017). Several nations with small populations are producing very high rates of CO_2 per capita, especially the top five: Qatar (43.9 metric tons), Trinidad and Tobago (37.3 metric tons), Kuwait (29 metric tons), Brunei Darussalam (24 metric tons), and Aruba (23 metric tons) (Millennium Development Goals Indicators, 2015).

Global temperature increases would have been more severe if it had not been for the protective effect of the oceans, which have been absorbing 90 percent of the carbon emissions released into the atmosphere (Katz, 2015). The heat buildup in the oceans has been tremendous--.5 watt to 1 watt of energy per square meter over the past ten years--amassing an energy that is roughly equivalent to five nuclear bombs exploding every second since 1990 (Katz, 2015). The huge quantities of CO_2 being released into the atmosphere and absorbed by the oceans have boosted the current rate of ocean acidification higher than at any time in the last 300 million years. Ocean acidification reduces seawater PH, carbonate ion concentration, and saturation states of calcium carbonate minerals that many marine organisms need to build their skeletons and shells (Miller & Armstrong, 2012). It also threatens the coral reefs of the world, as shellfish and corals need a calcium-rich environment; when the seawater becomes too acidic, they die off (Englander, 2013/14).

The Intergovernmental Panel on Climate Change (IPCC, 2017) was established in 1988 by the World Meteorological Organization (WMO) and United Nations Environment Programme (UNEP) to aid governments in developing climate-related policies by providing them with assessments of the scientific basis for climate change, its impacts and risks, and strategies for adaptation and mitigation. The IPCC's first assessment was published in 1990, its second assessment in 1995, its third assessment in 2001, its fourth assessment in 2007, and its fifth assessment in 2014; it is currently working on its sixth assessment, to be completed in 2019. In the Summary for Policymakers of the Fifth Assessment, the Working Group II of the IPCC opens with the following statement (IPCC, 2014, p. 3):

Human interference with the climate system is occurring, and climate change poses risks for humans and natural systems.

The Working Group II of the IPCC described the following observed impacts, vulnerability, and exposure:

- Recent changes in climate have impacted natural and human systems on all continents and across the oceans.
- In many regions, changing precipitation or melting snow and ice are altering hydrological systems, affecting water resources
- Many terrestrial, freshwater, and marine species have shifted their geographic ranges, seasonal activities, migrations, abundances, and species interactions in response to climate change
- Climate change has had more negative impacts on crop yields than it has had positive impacts
- There has been increased heat-related mortality and decreased cold-related mortality but the burden of ill-health from climate change is not well quantified
- Differences in vulnerability and exposure to climate change arise from multidimensional inequalities
- Impacts from climate-related extremes, such as heat waves, droughts, floods, cyclones, and wildfires, reveal significant vulnerability of human and ecosystems
- Climate-related hazards exacerbate the effects of other stressors, often negatively impacting livelihood, especially among those living in poverty
- Violent conflict increases vulnerability to climate change (IPCC, 2014).

The IPCC notes that until the mid-21st century, climate change will mostly exacerbate existing health problems. Thereafter, it is expected to result in more injuries, illnesses, and deaths from intense heat waves and fires, increasing numbers of people with malnutrition from diminished food production in poor regions, lost livelihood and labor productivity in vulnerable populations, and increased risks from foodborne, waterborne, and vector-borne diseases.

At the Earth Summit in Rio de Janeiro in 1992, for the first time the countries of the world acknowledged that climate change posed a serious global risk. Five years later, in Kyoto at the Parties of Conference-3 Summit, the Kyoto Protocol was adopted, limiting greenhouse emissions from 36 industrialized nations during the period 2008-2012. At the Copenhagen Climate Summit in 2009, it was agreed that 2 degrees Celsius should be the upper limit of temperature rise since the pre-industrial era, but no targets were put in place beyond what the Kyoto Protocol already stipulated. The Paris Agreement, adopted by 196 parties on December 12, 2015 in at the United Nations Framework Convention on Climate Change (UNFCCC) in Paris, was an historic moment in global efforts to slow down global warming. The main aim of the Paris Agreement is to limit global warming to a maximum of 2 degrees Celsius (3.6 degrees Fahrenheit) over the pre-industrial temperature by the year 2100 (United Nations, 2015). The Paris Agreement also stated that limiting global warming to a temperature rise of no more than 1.5 degrees Celsius (2.5 degrees Fahrenheit) over the pre-industrial temperature was an aspirational goal (United Nations, 2015). The average global temperature on Earth has increased by about 0.8 degrees Celsius (1.4 degrees Fahrenheit) since 1880, with two-thirds of the rise after 1975, at a rate of roughly 0.15-0.20 degrees Celsius per decade (NASA, 2017). Without international efforts to reduce carbon emissions, it is estimated that global temperatures will rise 3.5 degrees Celsius by the end of this century. Even with a 2 degree Celsius rise, there will be serious health consequences (see Chapter 2). The Paris Agreement assigned the IPCC the task of detailing the steps nations needed to take to achieve both the 2-degree limit and the 1.5 degree limit and to publish their recommendations by 2018. After the IPCC report is published, the nations who signed the Paris Agreement will have until 2020 to re-submit their pledges in accord with IPCC recommendations (Lyman, 2015). In November 2016, the UNFCCC met in Marrakesh, Morocco; it reaffirmed the goals of the Paris Agreement and established the Marrakech Partnership for Global Climate Action to help catalyze climate action by Parties and non-Party stakeholders in 2017-2020 (United Nations, 2016). While some progress has been made since 2015, much more needs to be done to reduce global warming, as is clear from a review of the key indicators of climate change.

KEY INDICATORS OF CLIMATE CHANGE

Global surface temperatures and arctic sea ice extent are two key indicators of climate change (U.S. EPA, 2012). Both had record-breaking years in 2016.

Average Global Surface Temperature

The year 2016 was confirmed as the hottest year on record for the world and for the United States; 2016 registered the highest global average surface temperature since record-keeping began in 1880, according to data reported by the World Meteorological Organization (WMO, 2017), NASA's Goddard Institute for Space Studies (NASA, 2017), the European Centre for Medium Range Weather Forecasts (ECMWF, 2017), and the US. National and Oceanic

Atmospheric Administration (Richter-Menge et al., 2016). The World Meteorological Organization is an agency of the United Nations based in Geneva, Switzerland with weather forecasting at more than 10,00 stations worldwide. The ECMWF is a research and weather forecasting service based in near London in the United Kingdom that comprises 34 European nations and provides datasets to members worldwide. The U.S. National and Oceanic Atmospheric Administration (NOAA) reports weekly, monthly and seasonal weather forecasts and climate predictions.

According to NASA (2017), the average global surface temperature was 1.78 degrees Fahrenheit (0.99 degrees Celsius) higher than the mid-20th century mean. NOAA (2017a) found that the average global surface temperature was 1.69 degrees Fahrenheit (.94 degrees Celsius) higher than the mid-20th century mean. Furthermore, both NASA and NOAA concurred that the 2016 warming was not an isolated event, but rather, part of a trend; in fact, 2014 and 2015 also experienced record-breaking hot temperatures. Only a small percentage of the 2015-16 warming (about .2 degrees Fahrenheit, or .12 degrees Celsius) could be attributed to the influence of El Nino (a warming of the ocean surface, or above-average sea surface temperatures, in the central and eastern tropical Pacific Ocean), which was in effect in 2015 and the first third of 2016 (NASA, 2017). NASA reports that most of the recorded warming of the Earth's surface temperatures occurred in the last 35 years, with 16 of the 17 hottest years on record reported in the period 2001-2016. NASA's findings are based on surface temperature measurements from 6,300 weather stations, ship- and buoy-based observations of sea surface temperatures, and measurements of temperatures at Antarctic research stations. The raw data are analyzed using an algorithm that considers the spacing of the weather stations as well as urban heating effects that could skew results (NASA, 2017).

Global temperature percentiles reported by the NOAA's National Centers for Environmental Information (NOAA, 2017b) appear in Figure 2.

Some areas of the globe are warming more quickly than others are. Most notably, the Arctic Program of the NOAA reported that in 2016 the arctic air temperatures were increasing at *double* the rate of the global temperature increase (Richter-Menge et al., 2016). Winter air temperatures were well above the previous record, with January temperatures 8 degrees Celsius above the norm in several locations (Richter-Menge et al., 2016).

Rising temperatures, and more severe and frequent heat waves related to climate change, are projected to increase heat-related illnesses and fatalities, especially among the elderly, people with asthma and other chronic medical conditions, and low-income populations without adequate access to air

Figure 2. Global temperature percentiles 2016
Source: NOAA (2017b).

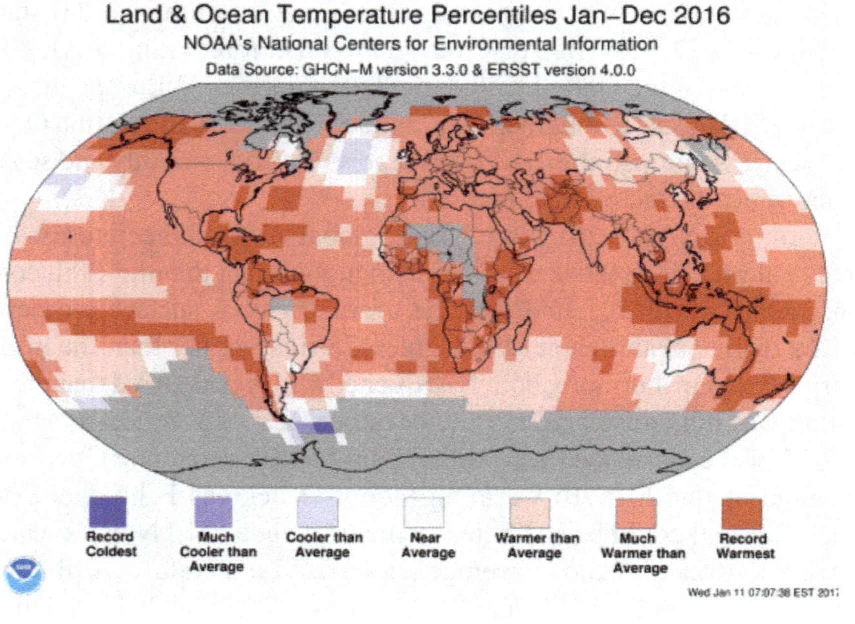

conditioning. Rising temperatures already have contributed to the increase of Salmonella prevalence in food; longer warm seasons and warmer winters increase the risk of food contamination and spoilage, and increase the number of pests and pest-transmitted diseases (see Chapter 2).

Extent of Arctic Sea Ice

NOAA (Richter-Menge et al., 2016) found that in 2016 the average Arctic sea ice extent was 3.92 million square miles, the smallest annual average since record-keeping began in 1979. Arctic sea ice extent is at its seasonal high (maximum extent) in March and at its seasonal low in September; that low point has been declining 13.4 percent per decade. The ice maximum extent reached a record low in 2016, following record-breaking surface temperatures. The average Antarctic sea ice extent for 2016 was 4.31 million square miles, the second smallest annual average since record-keeping began in 1979 (Richter-Menge et al., 2016).

Albedo is a measure of how well a surface reflects solar energy. It is between 0 and 1, with 1 being a perfect reflector and 0 being a perfect absorber, with no light reflected. The albedo of sea ice is much higher than the albedo of other earth surfaces, including the surrounding ocean. Typically, ocean albedo is approximately 0.06, while bare sea ice albedo varies from approximately 0.5 to 0.7; that means that the ocean reflects only 6 percent of the incoming solar radiation and absorbs the rest, while sea ice reflects 50 to 70 percent of the incoming energy (National Snow and Ice Data Center, 2017). The bright white surface of arctic sea ice acts like a giant reflector, sending solar energy back into space that otherwise would be absorbed into the ocean. The surface of snow is even more reflective than that of sea ice; when arctic ice is covered with snow, its surface reflects up to 90 percent of the solar energy. When the snow and ice melt, they are converted into dark seawater, which reflects only about 6 percent of the heat. This phenomenon causes millions of miles of the Earth's ocean surface to absorb 10-15 times more heat energy than it had when it was frozen ice covered with snow. The increased heat accelerates global warming and creates more water vapor, which, in turn, heats up the atmosphere and increases the rate of arctic ice melt (Englander, 2013/2014).

GLOBAL WARMING AND SEA LEVEL RISE

Global warming is causing sea levels to rise and greatly increasing coastal flooding risks (Union of Concerned Scientists, 2017). In his book, *High Tide on Main Street: Rising Sea Level and the Coming Coastal Crisis* (2013/2014), oceanographer John Englander noted that "the profound and permanent threat of sea level rise is barely appreciated" (p. viii). The first edition of his book came out just before Superstorm Sandy devastated New York in October 2012; the terrible aftermath of the flooding raised the public's consciousness about the reality of rising sea levels, at least in New York! Englander (2013/2014) warned:

As sea level rises, the shoreline will move hundreds, even thousands, of feet inland, destroying vast amounts of property, including most coastal communities…. The change will last for centuries, eventually reaching many tens of feet higher than now (p. viii).

Sea level changes are a function of the average long-term temperatures of the atmosphere and ocean. Over millions of years, sea levels and risen and fallen hundreds of feet. However, in the past 100 years, the rapid and enormous increase in carbon emissions from the burning of fossil fuels has created a human-made global warming. When fossil fuels—coal, oil, natural gas—are burned, carbon dioxide is released into the atmosphere and it acts like a blanket, trapping heat in the atmosphere. CO_2 levels and temperatures rise together. Some climate change skeptics attribute the CO_2 increases to volcanic eruptions. However, the emissions from volcanoes are minimal compared to the emissions from human burning of fossil fuels. The U.S, Geological Survey (Gerlach, 2011) reported that "On average, human activities put out in just three to five days the equivalent amount of carbon dioxide that volcanoes produce globally each year." Since 1880, when record keeping began, the global average temperature has increased 1.5 degrees Fahrenheit, with more than half of the increase in the past 35 years (Union of Concerned Scientists, 2017). The past three years (2014-2016) were the hottest on record, with 2016 experiencing the hottest average annual global temperature since record keeping began (NASA, 2017; NOAA, 2017a).

Sea level rises when land ice melts. The two biggest masses of land ice on the planet are in Greenland and Antarctica. The melting of the Greenland Ice Sheet would raise sea level 24 feet, while the melting of Antarctica would raise sea level 185 feet!

Figure 3 shows the trend in absolute sea level rise for Europe based on satellite measurements from 1992-2014.

Figure 3. Trends in absolute sea level 1992-2014
Source: European Environment Agency (EEA) (2017).

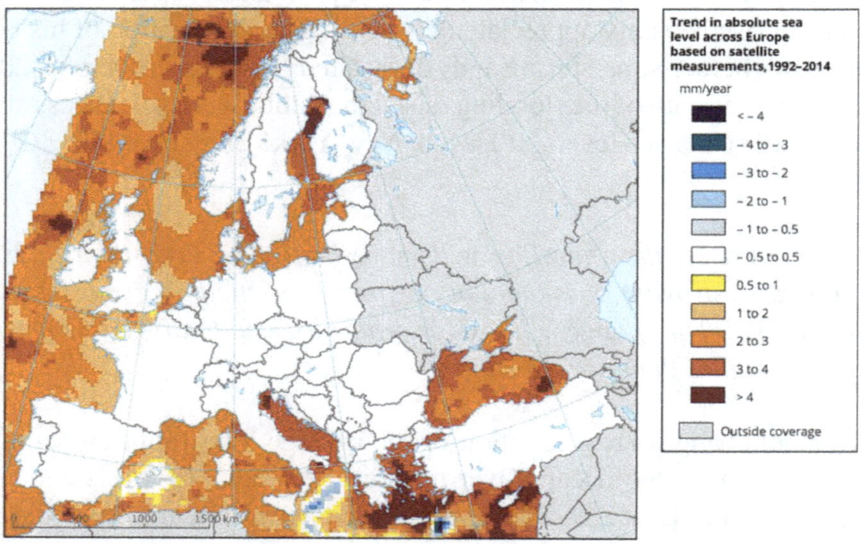

Absolute sea level rise does not take into account land rise. However, sea level rise relative to land along most European coasts is projected to be similar to the global mean, with the exception of the northern Baltic Sea and the northern Atlantic Coast, which are experiencing considerable land rise as a consequence of post-glacial rebound (EEA, 2016). The rise in global mean sea level was 19.5 cm (6.7 inches) from 1901 to 2015, at an average rate of 1.7 mm (.07 inches)/year. The rate of sea level rise accelerated after 1993, when satellite measurements became available, to approximately 3 mm (.12 inches)/year. Global mean sea level in 2015 was the highest yearly average over the record and ~70 mm (2.76 inches) higher than in 1993 (EEA, 2016).

In January 2017, the National Oceanographic and Atmospheric Association (NOAA) (Sweet et al., 2017) released an update of its findings on sea level rise and predictions for the future. According to the NOAA, Global mean sea level (GMSL) has increased by about 21 centimeters (cm)-24 cm (8-9 inches [in]) since 1880, with about 8 cm (3in) occurring since 1993 (Church & White, 2011; Hay et al., 2015; Nerem et al., 2010), and it is expected to continue to rise throughout the 21st century and beyond. The NOAA projected six scenarios-- Low, Intermediate-Low, Intermediate, Intermediate-High, High and Extreme, which correspond to GMSL rise of 0.3 m (11.8 in), 0.5 m (19.7 in), 1.0 m (39.4 in), 1.5 m (59.1 in), 2.0 m (78.7 in) and 2.5 m (98.4 in), respectively. The higher sea levels exacerbate the impacts of storm surge, high tides, and wave action (Theuerkauf et al., 2014). The six scenarios are shown in Table 1.

The Intermediate Scenario predicts a sea level rise of 1 meter by 2100, which would inundate many coastal cities around the world. The state that would be most affected in the United States is Florida, which has the most coastline and is flat, with most of the state at or slightly above sea level. Melting of the land ice on Greenland and Antarctica accounts for approximately 70

Table 1. Six scenarios of global mean sea level rise (GMSL) (meters)

GMSL	2010	2020	2030	2040	2050	2060	2070	2080	2090	2100	2150	2200
Low	0.03	0.06	0.09	0.13	0.16	0.19	0.22	0.25	0.28	0.30	0.37	0.39
Intermediate-Low	0.04	0.08	0.13	0.18	0.24	0.29	0.35	0.4	0.45	0.50	0.73	0.95
Intermediate	0.04	0.10	0.16	0.25	0.34	0.45	0.57	0.71	0.85	1.0	1.8	2.8
Intermediate-High	0.05	0.10	0.19	0.30	0.44	0.60	0.79	1.0	1.2	1.5	3.1	5.1
High	0.05	0.11	0.21	0.36	0.54	0.77	1.0	1.3	1.7	2.0	4.3	7.5
Extreme	0.04	0.11	0.24	0.41	0.63	0.90	1.2	1.6	2.0	2.5	5.5	9.7

Source: Adapted from Table 5, NOAA, 2017c.

percent of sea level rise (Hine et al., 2016), and the melting is accelerating. If the unstable Pine Island Glaciers in West Antarctica collapse and slide into the sea, the scenario will be much more serious than the Intermediate Scenario (Englander, 2013/2014).

Risk analysis and emergency preparedness planning, including an assessment about future changes in sea level, are of vital importance for states with coastal communities (White House, 2014). Some communities in low lying coastal areas have already taken steps to mitigate infrastructure to avoid flooded roadways. For example, Miami Beach has designed an upper tier for both street and sidewalk, elevated 2 ½ feet above the front doors of roadside businesses, and backed by a nearby pump house — what one city engineer called "the street of tomorrow" (Flechas & Staletovich, 2015). Miami Beach will install 60 pumps in all. The adaptation project – elevating roadways and sidewalks and installing pumps--is quite costly, estimated to ultimately run $400-$500 million. There is no roadmap, as this is an innovation, a new design that will keep Miami Beach's sidewalks and streets from flooding for about 40 years or so. Miami Beach leaders have acted with urgency, approving contracts on an emergency basis to fast-track the work (Flechas & Staletovich, 2015).

INFECTIOUS DISEASES AND FOOD POISONING

Climate change events such as natural disasters, flooding, droughts, and changes in precipitation have exacerbated the global problems of infectious diseases from water contamination and from infected mosquitoes, ticks, and flies; it has also increased the incidence of food poisoning. According to the World Health Organization (2017a), water-related diseases are the leading cause of disease and death worldwide, causing malnutrition from severe diarrhea and infecting millions with diseases such as cholera and typhoid. Most of the fatalities are in young children, the vast majority of whom die from exposure to water sources contaminated by raw sewage. The WHO/UNICEF Joint Monitoring Programme for Water Supply and Sanitation estimates that at least 1.8 billion people worldwide drinking water that is contaminated by feces. An even greater number drink water from systems that do not provide adequate protection against sanitary hazards. Examples of adequate protection against sanitary hazards are public taps or standpipes, tube wells or boreholes, protected dug wells, protected springs and rainwater collection (UN Water, 2014). Rising temperatures and sea levels, heavy rainfall and flooding contribute to water contamination by pollutants, agricultural runoff, and human and animal fecal matter.

In extremely poor regions in sub-Saharan Africa and Southeast Asia, where millions lack indoor plumbing or outdoor latrines, and open defecation is not uncommon, there is a severe problem with fecal contamination in the water supply, which is exacerbated by rising temperatures, heat waves, and droughts. According to NationMaster (2003-2017), which ranks countries on multiple variables affecting health and environmental safety, Somalia was last on availability of clean drinking water; unsurprisingly, it also had the highest rate of cholera (waterborne illness) of any nation. Somalia is a nation mired in poverty, facing multiple environmental problems including famine, deforestation, overgrazing, soil erosion, desertification; the life expectancy at birth is 50.4 years. The United States ranks 62nd out of 194 nations on availability of clean drinking water; it is a wealthy nation with significant pockets of poverty—15.1 percent of the population was below the poverty line in 2011 (NationMaster, 2003-2017).

A vector is an insect, tick or other organism that transmits a disease from a plant or animal to an animal by biting it. Vectors include mosquitoes, ticks, triatomine bug, sandflies, and blackflies. Rising temperatures due to climate change affect the survival and reproduction rates of vectors, as well as vector habitat suitability, distribution, and abundance (Semenza & Menne, 2009). For example, warmer temperatures expand tick activity, leading to increased exposure to the tick that transmits Lyme disease and, hence, an increased prevalence of Lyme disease in animals and humans. According to the World Health Organization (WHO, 2016b), there are more than 1 billion cases of vector-borne diseases annually, resulting in more than 1 million deaths per year. Vector-borne diseases comprise more than 17 percent of all infectious diseases (WHO, 2016b).

Mosquitoes are responsible for the most vector-borne diseases, including chikungunya, dengue fever, Japanese encephalitis, Lymphatic filariasis, malaria, Rift Valley fever, yellow fever, West Nile fever, and zika. Dengue fever is the most prevalent of these diseases; the World Health Organization (WHO, 2017b) reports that it has spread dramatically in recent years, with a reported 3.9 billion people in 128 countries at risk of contracting dengue fever. Lymphatic filariasis (elephantiasis) is estimated to threaten 947 million people in 54 countries worldwide.

Sandflies are responsible for Leishmaniasis and sandfly fever (phlebotomus fever), while tse-tse flies are responsible for sleeping sickness (African trypanosomiasis). Fleas transmit rickettsiosis (spotted fever and typhus) and plague by infecting rodents and other animals who, in turn, infect humans; in the fourteenth century, the bubonic plague spread by flea-infected rats and wiped out more than half of Europe's population. Today plague is still widely distributed in the tropics and subtropics (WHO, 2017b). Other vector-borne diseases include river blindness (onchocerciasis) spread by black flies and schistosomiasis (bilharziasis). About 90 percent of the cases of river blindness occur in Africa; in some West African communities, approximately 50 percent of men over 40 have been blinded by this disease, which is treatable. It is the second most common cause of blindness worldwide, after trachoma. Management of river blindness involves both vector control and treatment of river blindness with antibiotics before the damage becomes irreversible (WHO, 2016b).

Warmer temperatures cause food to spoil more rapidly. The World Health Organization reported that, worldwide, 31 hazards caused 600 million foodborne illnesses and 420,000 deaths worldwide in 2010. Africa had the heaviest burden of foodborne diseases, followed by Southeast Asia and the Eastern Mediterranean (WHO, 2015). *Campylobacter* and *Salmonella* were the two most common bacteria responsible for foodborne diseases. Even in the United States, which has one of the safest food supplies, *Campylobacter* bacteria (which causes food poisoning) was found on 47 percent of raw chicken samples tested in grocery stores in 2011 (CDC, 2014). Proper hygiene when handling raw poultry, meat, and eggs is extremely important to prevent food spoilage, but it is very difficult to do in regions that lack plumbing, have very limited water supplies due to drought, and have no refrigeration (see Chapter 2 for discussion on prevention of food spoilage).

UNITED STATES TRENDS AND IMPACTS

In *Climate Change: Impacts in the United States: The Fourth National Climate Assessment,* Balbus et al. (2016) have identified the following nine regional climate trends in the United States:

- Rising temperatures, especially in North and West;
- Rising sea levels, especially along Mid-Atlantic and Gulf Coast;
- More extreme and frequent North Atlantic hurricanes;
- Extreme precipitation in the Midwest and Northeast;
- Increasing floods in the Midwest and Northeast;
- More frequent and intense winter storms in the Plains;

- Increasing wildfires in the West;
- More frequent and intense heat waves, especially in the West;
- More persistent droughts in the Southwest.

Rising Temperatures

The average temperature in the United States has increased by 1.3 degrees Fahrenheit to 1.9 degrees Fahrenheit since record keeping began in 1895; most of the increase has been since the 1970's. The last decade was the nation's warmest on record, and temperatures are expected to continue to rise. The frost-free season has been getting longer since the 1980s, especially in the West, affecting ecosystems and agriculture. Extreme temperatures are projected to increase at both ends of the spectrum (Balbus et al., 2016).

Rising Sea Level

Nearly 40 percent of the U.S. population lives in relatively densely populated coastal areas, where rising sea level has contributed to flooding, shoreline erosion, and hazards from storms. The higher sea levels push storm surges farther inland, which results in much more frequent nuisance flooding (flooding that leads to public inconveniences such as road closures) (NOAA, 2016). The higher sea levels have led to street flooding in Miami Beach, Fort Lauderdale, Hollywood, and other coastal communities during King Tides (biannual tides when the sun, moon and earth align) in Florida.

More Extreme and Frequent North Atlantic Hurricanes

The year 2016 had above average North Atlantic Hurricane activity, 140 percent of normal Accumulated Cyclone Energy (ACE), with 15 storms and 7 hurricanes. Matthew was the first Category 5 hurricane in the North Atlantic since Hurricane Felix in 2007. Matthew wreaked major damage on Haiti, Cuba, and the Bahamas, and part of the Southeastern United States, with more than 1000 fatalities (Voice of America, 2016). Higher sea levels add to the impact of hurricanes, causing extreme storm surges and flooding. Adequate emergency preparedness, including education about whom to call and what to do when a disaster strikes, will mitigate loss of life and property. Cultivating social cohesion fosters climate resilience, a key factor in recovery from a hurricane (see Chapter 7).

The National Weather Service (NWS) (Uccellini, 2016) urges Americans to make the United States a Weather-Ready Nation by building community resiliency in the face of increasing vulnerability to extreme weather events. The NWS is working to create a Unified Global Coupled System (UGCS) that would couple Atmosphere, Ocean, Land-Hydrology, Sea-Ice, Wave, and Aerosol (dust, sea salt, sulfate, and carbonaceous aerosols) components. UGCS would unify Weather (GFS), Sub-seasonal (GEFS), and Seasonal (CFS) forecasts and analyses (Uccellini, 2016).

Extreme Precipitation and Floods in the Midwest and Northeast

In the western Great Lakes region, as much as half of annual total precipitation falls during 10 days of the year. From 1958 to 2012, the amount of precipitation that falls during the most intense 1 percent of precipitation events increased by 37 percent in the Midwest and 71 percent in the Northeast (Walsh et al., 2014). Extreme precipitation leads to stormwater management problems, as it overloads sewage systems and water treatment facilities, and increases the risk for waterborne disease outbreaks from untreated sewage discharged into surface water (Karl et al., 2009). Rising sea surface temperatures, rising sea levels, extreme precipitations, and floods affect the growth, survival, spread, and toxicity of bacteria and other agents of water-related illnesses. Heavy rains and flooding cause excessive nitrate and phosphorus from fertilizer runoff to enter the freshwater and seawater, resulting in eutrophication, a type of water pollution that gives rise to toxic algae blooms (NOAA, 2008). This problem is particularly severe in the Midwest and Northeast (National Wildlife Federation, 2017). The main sources of water contamination are feces (human and animal) and agricultural fertilizers. In the United States, fecal contamination of water—the major global source of human infection by water-borne disease--occurs when the water infrastructure is overwhelmed by heavy rains and runoff (Trntanj et al., 2016).

Heavy downpours and flooding associated with climate change may cause damp spots inside buildings and homes. Mold spores can grow on damp spots on the wall or ceiling (including water leaks inside walls or between the ceiling and the floor of the room above) and compromise indoor air quality (House et al., 2016). Molds produce allergens (substances that can cause allergic reactions), irritants, and in some cases, potentially toxic substances (mycotoxins) (U.S. EPA, 2017). Sensitive individuals who inhale or touch mold may have allergic reactions such as sneezing, runny nose, red eyes, and skin rash (dermatitis). People with asthma who are allergic to mold may have an asthma attack after exposure to mold. Even those who are not allergic to mold may experience irritation of the eyes, skin, nose, throat, and lungs from exposure to mold (U.S. EPA, 2017).

Extreme Winter Storms in the North

Roughly twice as many extreme U.S. snowstorms occurred in the second half of the 20th century than did in the first. Warmer air over the Atlantic Ocean causes additional moisture to gather into storms, which then travel inland and dump unusually large amounts of snow in the North. In February 2010, two back-to-back blizzards broke records in the Mid-Atlantic for most snowfall, and were dubbed Snowmaggeddon (Oskin, 2013). On March 14, 2017, Snowstorm Stella pounded the Northeast, closing schools and airports all down the Northeast corridor. While two feet or more accumulated in a number of places, the record for most snow goes to Bolton Valley, Vermont, where the snowfall totalled 58 inches (Abramson & Bachor, 2017).

Increasing Wildfires, More Severe Heat Waves, and More Persistent Droughts

California has just come out of a severe five-year drought, which dried out trees and sparked numerous wildfires. In March 2015, during the drought, an intense heat wave broke temperature records from Los Angeles all the way north to Sacramento (Rocha, 2015).

Higher temperatures, decreased precipitation, and wildfires lead to increased levels of ozone, particulate matter, and pollen counts in outdoor air, increasing the risk of cardiovascular and respiratory illness and death. In the United States, the American Lung Association's State of the Air 2016 found that, even with some improvement in levels of particle pollution and ozone, 166 million Americans still lived in counties where they were exposed to unhealthy levels of these pollutants (American Lung Association, 2016). Data were missing from three states, including from two of the most polluted cities (Los Angeles and Chicago) from prior studies. In addition, the report "provides evidence that a changing climate will make it harder to protect human health" (American Lung Association, 2016, p. 4). (See Chapter 2 for discussion of heat-related illnesses.)

GLOBAL TRENDS AND IMPACTS

Europe

In Europe, global warming has created rising sea levels along the coastlines; temperatures have been rising, with much larger than global average increases in Northern Europe and the Arctic. There has been a decrease in Arctic sea ice, a decrease in the Greenland ice sheet and permafrost, and intensified shipping and exploitation of oil and gas resources in the Arctic, with an increasing risk of biodiversity loss (European Environment Agency, 2016). Sea surface temperatures have risen, as has the ocean's acidity. In the Mediterranean, there has been a decrease in annual precipitation and river flow, an increasing risk of desertification and biodiversity loss, an increasing water demand for agriculture and decrease in crop yields, increase in mortality from heat waves, expansion of vector habitats, and a decrease in hydropower potential. In Central and Eastern Europe, summer precipitation has decreased and the risk of forest fires has increased, as have the temperatures on land and sea (European Environment Agency, 2016).

To make Europe more climate resilient, in 2013 the European Commission (EC) adopted *An EU Strategy on Adaptation to Climate Change* (EC, 2013), which supports climate action toward adaptation, as well as research and information-sharing. It also supports "mainstreaming," that is, integrating concerns about adaptation into existing sectoral EU policies (e.g., agriculture) (European Environment Agency, 2016). In addition, adaptation is also considered in other EU initiatives such as *Europe 2020 — Europe's growth strategy,* the EU's ten-year plan launched in 2010 to create the conditions for smart, sustainable and inclusive growth (*Europe 2020*, 2016), and in the *Resource-Efficient Europe* flagship initiative, one of seven put forth in *Europe 2020.*

Asia

China is the fourth largest country in the world in terms of land mass; it is the most populous country in the world, with 1.371 billion people, a little less than one-fifth of the world's population (World Bank, 2016). The country's weather system is dominated by monsoons, which bring cold, dry air down from Siberia during the winter, and warm, moist air up from the Indian subcontinent and South China Sea during the summer. Since 1978, one-quarter to one-third of China's farmland has had reduced crop yields due to floods, droughts, wind, hail, typhoons, cold spells, freezes, and snow (Xinhua, 2012). The government estimates that extreme weather events cost the country 1 percent to 3 percent of its gross domestic product (GDP) each year (see Xinhua 2010). China's exposure to the impacts of extreme weather events is exacerbated by its lack of restrictive zoning codes and of higher insurance premiums to discourage settlement and economic activities in the most hazardous areas.

In China, droughts are increasing in the North, Northeast, and Northwest; flooding is increasing in the middle and lower Yangtze and Pearl River basins, extreme precipitation is projected in some regions, the melting of glaciers is affecting river flows downstream, and heat waves are becoming more frequent (Sall, 2013). China is experiencing severe air and water pollution, as well as desertification in the Northwest. The World Health Organization tested the water at a major reservoir that supplies drinking water to Beijing and found that the water had lead levels 20 times the maximum limit, despite a $850 million government effort to clean up the water pollution (Ludacer, 2015). This cleanup effort was initiated in 2013, after more than 2,000 dead pigs were found floating in Shanghai River, a main water source for the city's 23 million people (Spector, 2013). In 2015 there was a severe blue algae outbreak in China's fifth largest freshwater lake, Chaohu Lake, and a "green tide" of algae 35,000 square km across in waters off China's eastern coast (Mengjie, 2015).

India, the world's second most populous country, with 1.31 billion people (World Bank, 2016), has the most polluted air in the world. Air pollution causes 1.1 million deaths per year. Weak environmental regulations force Indians to take their complaints to India's Environmental Court, the Green Tribunal, but it is not able to enforce its rulings (Anand, 2017). India's water is even worse; approximately 80 percent of sewage is untreated and flows directly into the rivers, polluting the main sources of drinking water. The fecal contamination in water causes outbreaks of water-borne diseases such as cholera (Narayanan, 2015). One-half of India's population – 620 million people – lack latrines and defecate outside. When children are exposed to open defecation or do not have a clean water supply, they ingest bacteria, viruses, fungi, or parasites that cause intestinal infection. Stunting—having a smaller than normal stature due to malnutrition—may result, along with anemia and poor early childhood development. India has an estimated 61.7 million stunted children—one quarter of the world's total.

The dire conditions of sanitation are exacerbated by extreme heat waves. In May 2016, India experienced an extreme heat wave, which resulted in the deaths of an estimated 700 people (BBC News, 2016). The heat wave hit much of northern India and lasted for weeks. On May 19, in Phalodi in the state of Rajasthan, the thermometer read 51 degrees Celsius (123.8 degrees Fahrenheit), the hottest temperature ever recorded in India (BBC News, 2016).

In Indonesia, the widespread burning of peatlands has created a dense haze of pollution over Sumatra and Borneo, and gas masks have been distributed to protect the population from particulate matter in the air. The thousands of fires are threatening to drive endangered species, such as orangutans, rhinoceros, tigers and elephants, into extinction (Oelrichs, 2015). Rapid deforestation, overexploitation of resources, peat fires, floods and landslides have threatened the livelihoods of those living in forest and coastal areas. Approximately 40 percent of Indonesia's population lives in areas at risk from multiple hazards from disasters and climate risk (Dilley et al., 2005).

Africa

In sub-Saharan Africa, severe droughts have created famine conditions in Somalia, Kenya, and Ethiopia. Malnutrition contributes to more than one-third of all child deaths worldwide (WHO, 2017a). The droughts have caused food shortages and deaths of both humans and livestock. Overgrazing and burning of biomass are two human activities that contribute to drought conditions and resulting food shortages and starvation. Burning is done by herders to promote new grass growth and by farmers to remove unwanted biomass after the harvest season. Both burning biomass and overgrazing dry out the soil and disrupt the hydrological system, including rainfall patterns (Gaworecki, 2017). A recent study (Ichoku et al., 2016) found that the fires tended to suppress precipitation in drought-ravaged sub-Saharan Africa.

In addition, biomass burning on cookstoves is used in the household for heating and cooking by 81 percent of the population. The cookstoves create severe indoor air pollution, which poses serious health hazards to women. They also contribute to global warming by emitting black carbon (Cho, 2016). Sub-Saharan Africa and developing countries in Asia and Latin America emit more than ¾ of global black carbon emissions, from the burning of wood-based biomass (wood and charcoal) (Cho, 2016). The development and distribution of cleaner, more efficient cookstoves is one strategy to reduce the harm of the biomass burning (AFREA, 2011).

The Middle East

In the Middle East, climate change has been increasing the intensity and persistence of droughts, leading to water management problems. The drought problem has been exacerbated by scorching heat. In July 2016, Iran and Kuwait experienced severe heat waves, with an unearthly heat index of 73 degrees Celsius (163 degrees Fahrenheit) in Bandar Mahshahr, Iran on Friday, July 31, and recorded temperatures of 51 degrees Celsius (123.8 degrees Fahrenheit) in Baghdad, Iran and 50-52 degrees Celsius (122-125.6 degrees Fahrenheit) north of Kuwait City (Omar, 2015).

In Israel, an arid country that is 60 percent desert, a water revolution that utilizes innovative technologies has turned the nation into a world leader in water management. Israel reuses 86 percent of its wastewater for agriculture; farmers use drip irrigation, invented by an Israeli engineer, which uses much less water than previous methods and produces bigger crops. In addition, Israel has a centralized water control that regulates all water so as to make both residential and business users as conservative as possible. In addition, Israel uses desalination plants to provide 25 percent of the nation's water supply and 80 percent of household water (Distel, 2015).

Australia

Australia has a population of only 23.78 million (World Bank, 2016) yet emits as much CO_2 as Indonesia, which has 200 million people (Bourke, 2015). Its per capita CO_2 emissions were 16.2 metric tons in 2011, just slightly below the United States per capita CO_2 emissions of 16.8 metric tons (Millennium Development Goals, 2015). While world CO_2 emissions levelled off in 2015 (IEA, 2017), Australia's GHG emissions increased by nearly 1 percent, pointing to its urgent need to invest in renewable energy resources and put a moratorium on new coal mines (Bourke, 2015). Australia has been experiencing rising temperatures and intense heat waves. By 2030, temperatures are projected to rise by about 1 degree Celsius, with a greater increase inland than in coastal areas, and with more days of extreme heat (temperatures above 35 degrees Celsius) (Alexander & Arblaster, 2009; National Climate Centre, 2009). Rainfall is projected to decrease in the South during the winter, in the South and East during the spring, and on the West coast during the autumn. Although there are likely to be many more dry days, when it rains, the precipitation will be intense. Drought and wildfires are expected to increase. Projections are for a significant loss of biodiversity in the Great Barrier Reef and Queensland Wet Tropics; also at risk are the Kakadu wetlands, Southwest Australia, sub-Antarctic islands and the alpine areas (Global Greenhouse Warming, 2017).

Antarctica

Antarctica contains the greatest mass of land ice on the planet. There are unmistakable signs that it is melting at an accelerating rate. Photos taken in 1975 and 2013 show significant reduction of land ice (Englander, 2017). If just 3 percent of the land ice in Antarctica melts, the sea level will rise five feet and world's coastlines will be flooded (Englander, 2017). West Antarctica, which contains roughly one-third of the continent's land mass, is unstable; its Pine Island Glacier is melting and breaking apart from the inside out (Pandey, 2016). In 2015, a 225-square-mile iceberg broke off from the glacier. Analysis of satellite images showed that the rift that caused the break started nearly two years earlier in the center of the ice shelf (Pandey, 2016).

Global Hotspots for Natural Disaster

Natural Disaster Hotspots: A Global Risk Analysis (Dilley et al., 2005) identified the geographic areas at highest risk of disasters from earthquakes, volcanoes, landslides, floods, drought, and cyclones. Three indexes were developed—mortality risk for global gridded population, risk of total economic loss, and risk of economic loss as a percentage of GDP.

Key findings from the global risk analysis are summarized below:

- 3.4 billion people (more than 50 percent of the world's population) are relatively highly exposed to one hazard.
- 790 million people are relatively exposed to at least two hazards.
- 105 million are relatively highly exposed to at least three hazards.
- Geophysical hazards—earthquakes and volcanoes—cluster around fault lines in mountainous areas.
- Hydro-meteorological hazards—floods, cyclones, and landslides—strongly affect the eastern coastal regions of the major continents as well as some interior regions of North and South America, Europe, and Asia.
- Drought is widely dispersed among the arid semi-tropics.

- Regions subject to both geophysical and hydro-meteorological hazards are primarily in East and South Asia and in Central America and Western South America.

The hazards can interact with even more dire consequences, such as a cyclone and flooding triggering a landslide, or earthquakes damaging dams and reservoirs needed for drought and flood protection. The top country exposed to three or more hazards was Taiwan, China, followed by Costa Rica, Vanuatu, and the Philippines (Dilley et al., 2005). The risk of relatively high mortality from multiple hazards for the top hotspots was very high—nine countries had a risk of 90 percent or more, and two had a risk of 89 percent-90 percent. Many of the disaster hotspots had higher-than-average densities of GDP, leading to a relatively high degree of exposure of economically productive areas (Dilley et al., 2005).

Disaster risk analysis is a technological tool that should be used by global stakeholders in disaster management to prioritize and invest in emergency preparedness and sustainable development in high-risk areas.

HEALTH EDUCATION AND TECHNOLOGY

Health education and technology are key components of planning for a sustainable future. To change the mindset of a nation from fossil-fueled consumerism to green sustainability requires knowledge and persistence. The message needs to be sent over multiple channels, over time. A tipping point seems to have been reached in some countries. Germany made a big move in the direction of reducing carbon emissions by declaring that it would produce and register only electric vehicles (EVs) from 2030 on. The EVs have to be "sold" to the public via a social marketing campaign that provides economic and social incentives for purchasers of EVs. The economic incentives are already underway; it may take a few years to make EVs more popular than traditional vehicles. The German EV campaign is using DOI Theory to reach out to early adopters (opinion leaders who like to try innovations), who will convince the majority to try the innovation.

Educational materials on the health impacts of climate change need to be developed not only in English but in languages that immigrant groups speak. The effort needs to include schoolchildren. Community toolkits have been developed to empower communities to work together to improve their emergency preparedness for climate change events. This book gives examples of such initiatives. Legislators need to be informed that the public wants policies that incentivize technology-based solutions to keep communities safe through heat waves and extreme weather events. Change is already taking place, but the pace needs to be accelerated. You can be part of the solution!

CONCLUSION

Clearly, global warming cannot be stopped; even if all fossil fuel burning ceased immediately, the atmospheric concentration of CO_2 is so high that warming trend would continue for a long time. However, it is possible to slow down global warming, through a worldwide commitment to switch to renewable, clean energy sources. It is also imperative to help people adapt to climate change. The Paris Agreement in 2015 represented a tipping point in global awareness of the urgent need to take action in reducing GHG emissions and working on sustainable development. Subsequent chapters will address the use of innovative technologies and initiatives to address specific health impacts of climate change, with the overarching goal of achieving environmental justice.

REFERENCES

Abramson, A., & Bachor, K. (2017). Winter storm Stella. *Time*. Retrieved from http://time.com/4700759/storm-stella-snow-blizzard-photos/

Africa Renewable Energy Access Program (AFREA). (2011). *Wood-based biomass energy development for sub-Saharan Africa*. Washington, D.C.: The World Bank.

Alexander, L., & Arblaster, J. M. (2009). Assessing trends in observed and modelled climate extremes over Australia in relation to future projections. *International Journal of Climatology*, *29*(3), 417–435. doi:10.1002/joc.1730

American Lung Association. (2016). State of the Air 2016. ALA: Chicago, IL. Retrieved from http://www.lung.org/assets/documents/healthy-air/state-of-the-air/sota-2016-full.pdf

Anand, G. (2017). India's air pollution rivals China's as world's deadliest. The New York Times, February 14, 2017. New York: The New York Times Company. Retrieved from https://www.nytimes.com/2017/02/14/world/asia/indias-air-pollution-rivals-china-as-worlds-deadliest.html

Balbus, J., A., Crimmins, J.L., Gamble, D.R., Easterling, K.E., Kunkel, S., Saha, &. Sarofim, M.C. (2016). Introduction: Climate Change and Human Health. In *The Impacts of Climate Change on Human Health in the United States: A Scientific Assessment* (Ch. 1, pp. 25–42). Washington, DC: U.S. Global Change Research Program. doi:10.7930/J0VX0DFW

Boden, T., Andres, B., & Marland, G. (2017). Ranking of the world's countries by 2014 total CO2 emissions from fossil-fuel burning, cement production, and gas flaring. Oak Ridge, UT: Carbon Dioxide Information Analysis Center. Retrieved from http://cdiac.ornl.gov/trends/emis/top2014.tot

Bourke, L. (2015). Australia's carbon emissions jump in 2015. The Sydney Morning Herald, December 26, 2015. Sydney, Australia: Fairfax Syndication. Retrieved from http://www.smh.com.au/federal-politics/political-news/australias-carbon-emissions-jump-in-2015--20151225-glv017.html

British Broadcasting Channel (BBC). (2016). India records its hottest day ever. Retrieved from http://www.bbc.com/news/20039682

Centers for Disease Control and Prevention. (2014). Campylobacter. Atlanta, GA: US Department of Health and Social Services. Retrieved from https://www.cdc.gov/foodsafety/diseases/campylobacter/index/html

Centers for Disease Control and Prevention. (2015). Food Safety: Foodborne Germs and Illnesses. Atlanta, GA: Centers for Disease Control and Prevention. Retrieved from http://www.cdc.gov/foodsafety/foodborne-germs.html

Cho, R. (2016). *The Damaging Effects of Black Carbon. State of the Planet. Earth Institute*. New York: Columbia University.

Church, J. A., & White, N. J. (2011). Sea-level rise from the late 19th to the early 21st century. *Surveys in Geophysics*, *32*(4-5), 585–602. doi:10.1007/s10712-011-9119-1

Dilley, M., Chen, R. S., Deichmann, U., Lerner-Lam, A. L., Arnold, M., ... (2005). Natural Disaster Hotspots: A Global Risk Analysis. Washington, D.C.: World Bank. doi:10.1596/0-8213-5930-4

Distel, O. (2015). Israel's water revolution and its overflow for the world. Times of Israel Blog, June 27, 2016. Retrieved from http://blogs.timesofisrael.com/israels-water- revolution-and-its-overflow-for-the-world/

Englander, J. (2013/2014). *High Tide on Main Street: Rising Sea Level and the Coming Coastal Crisis* (2nd ed.). Boca Raton, FL: The Science Bookshelf.

Englander, J. (2017). Antarctica wake up call – my 4 expedition takeaways. Retrieved February 27, 2017 from http://www.johnenglander.net/sea-level-rise-blog/antarctic-wake-call-4-expedition-takeaways/

European Centre for Medium Range Weather Forecasts (ECMWF). (2017). 2016 was warmest year yet, ECMWF data show. Retrieved from http://www.ecmwf.int/en/about/media-centre/news/2017/2016-was-warmest-year-yet-ecmwf-data-show

European Commission. (2016). Europe 2020.

European Commission (EC). (2013). Communication from the Commission to the European Parliament, the Council, the European Economic and Social Committee and the Committee of the Region — An EU Strategy on adaptation to climate change COM 216 final. Brussels, Belgium: EC.

European Commission (EC). (2017). Climate action. Brussels, Belgium: EC. Retrieved from https://ec.europa.eu/clima/change/causes_en

European Environment Agency (EEA). (2016). Climate change impacts and adaptation. Copenhagen, Denmark: EEA. Retrieved from http://www.eea.europa.eu/soer-2015/europe/climate-change-impacts-and-adaptation#note4

European Environmental Agency (EEA). (2017). Trend in absolute sea level of Europe based on satellite measurements, 1992-2013. Copenhagen, Denmark: EEA. Retrieved from http://www.eea.europa.eu/data-and-maps/figures/sea-level-changes-in-europe-october-1992-may-1

Flechas, J., & Staletovich, J. (2015, October 23). Miami Beach's battle to stem rising tides. *Miami Herald*. Retrieved from http://www.miamiherald.com/news/local/community/miami-dade/miami-beach/article41141856.html

Fourier, J. (1824). Remarques generales sur les temperatures du globe terrestre et des espaces planetaires. *Ann. Chim. Phys.*, *27*, 136-167.

Gaworecki, M. (2017). NASA scientists find connection between fires and droughts in sub- Saharan Africa. *Mongabay*. Retrieved from https://news.mongabay.com/2017/01/nasa-scientists-find-connection-between-fires-and-droughts-in-sub-saharan-Africa/

Gerlach, T. (2011). Volcanic versus anthropogenic carbon dioxide. *Eos Trans. AGU*. Retrieved from http://www.agu.org/pubs/cos-news/supplements/2011/gerlach_92_24.shtml

Global Assessment Report on Disaster Risk Reduction. (2015). Making Development Sustainable: The Future of Disaster Risk Management. United Nations. Retrieved from http://www.preventionweb.net/english/hyogo/gar/2015/en/gar-pdf/GAR2015_EN.pdf

Global Greenhouse Warming. (2017). Climate change in Australia and New Zealand. Retrieved from http://www.global-greenhouse-warming.com/climate-change-in-Australia-and-New-Zealand.html

Hay, C. C., Morrow, E., Kopp, R. E., & Mitrovica, J. X. (2015). Probabilistic reanalysis of twentieth century sea-level rise. *Nature*, *517*(7535), 481–484. doi:10.1038/nature14093 PMID:25629092

Hine, A. C., Chambers, D. P., Clayton, T. D., Hafen, M. R., & Mitchum, G. T. (2016). *Sea Level Rise in Florida*. Gainesville, FL: University Press of Florida.

House, J., Brovkin, V., Betts, R., & Constanza, R. …. Scholes, M. (2016). Chapter 13: Climate and Air Quality. In: *The Impact of climate change on human health in the United States: a scientific assessment*. Washington, D.C.: U.S. Global Change Research Program. Retrieved from https://health2016.globalchange.gov/

Ichoku, C., Ellison, L. T., Willmot, K. E., Matsui, T., Dezfuli, A. K., Gatebe, C. K., & Okonkwo, C. et al. (2016). Biomass burning, land-cover change, and the hydrological cycle in Northern sub-Saharan Africa. *Environmental Research Letters*, *11*(9), 095005. doi:10.1088/1748-9326/11/9/095005

IPCC. (2014). Summary for policymakers. In C.B. Field, V.R. Barros, D.J. Dokken et al., (Eds.), Climate Change 2014. Cambridge, UK: Cambridge University Press.

IPCC. (2017). IPCC Fact Sheet: What is the IPCC? Retrieved from https://www.ipcc.ch/news_and_events/docs/factsheets/FS_what_ipcc.pdf

Karl, T. R., Melillo, J. M., & Peterson, T. C. (2009). Global Climate Change Impacts in the United States. Cambridge, MA: Cambridge University Press. Retrieved from http://downloads.globalchange.gov/usimpacts/pdfs/climate-impacts-report.pdf

Katz, C. (2015). How Long Can Oceans Continue to Absorb Earth's Excess Heat? Yale Environment 360. New Haven, CT: Yale School of Forestry and Environmental Studies. Retrieved from http://e360.yale.edu/features/how_long_can_oceans_continue_to_absorb_earths_excess_heat

Lang, K. R. (2010). Global warming. In NASA's Cosmos. Medford, MA: Tufts University. Retrieved from at http://ase.tufts.edu/cosmos/view_chapter.asp?id=21&page=1

Ludacer, R. (2015, August 15). If you think China's air is bad, you should see the water. *Business Insider*. Retrieved from http://www.businessinsider.com/china-water-pollution-2015-8

Lyman, E. (2015). 5 takeaways from the climate deal. *USA Today*. Retrieved from http://www.usatoday.com/story/news/2015/12/12/paris-agreement-5-key-takeaways-historic-climate-deal/77207284/

Maslin, M. (2004). *Global Warming, a very short introduction*. Oxford, UK: Oxford University Press.

McKibben, B. (2013). Do the Math (the movie). New York: 350.org & PF Pictures.

Mengjie. (2015, June 4). Algae bloom expands in China's Yellow Sea. English.news.cn., Qingdao, Shandong, China. Retrieved from http://news.xinhuanet.com/english/2015-06/04/c_134297379.htm

Millennium Development Goal Indicators. (2015). New York: United Nations. Retrieved from https://mdgs.un.org/unsd/mdg/SeriesDetail.aspx?srid=751

Miller, J., & Armstrong, T. (2012). Study Finds Ocean Acidification Rate is Highest in 300 Million Years, CO2 is Culprit. Washington, D.C.: U.S. Global Change Research Program. Retrieved from http://www.globalchange.gov/news/study-finds-ocean-acidification-rate-highest-300-million-years-co2-culprit

Narayanan, N. (2015). Horrifying fact: Almost all India's water is contaminated by sewage. Retrieved from https://scroll.in/article/737981/horrifying-fact-almost-all-indias-water-is-contaminated-by-sewage

NASA. (2017). NASA, NOAA data show 2016 hottest year on record. Retrieved from https://www.nasa.gov/press-release/nasa-noaa-data-show-2016-warmest-year-on-record-globally

National Climate Centre. (2009). The exceptional January-February 2009 heatwave in south-eastern Australia. Special Climate Statement 17, Bureau of Meteorology, (updated 12 February). Retrieved from www.bom.gov.au/climate/current/statements/scs17c.pdf

National Oceanic and Atmospheric Administration. (NOAA). (2016). What is nuisance flooding? Washington, D.C.: US Department of Commerce. Retrieved from http://oceanservice.noaa.gov/facts/nuisance-flooding.html

National Oceanic and Atmospheric Administration (NOAA). (2008). Nutrient pollution – Eutrophication. Washington, D.C.: US Department of Commerce. Retrieved from http://oceanservice.noaa.gov/education/kits/estuaries/media/supp_estuar09b_eutro.html

National Oceanic and Atmospheric Administration (NOAA). (2017a). 2016 marks the third consecutive year of record warmth for the globe. Washington, D.C.: The Department of Commerce. Retrieved from http://www.noaa.gov/stories/2016-marks-three-consecutive-years-of-record-warmth-for-globe

National Oceanic and Atmospheric Administration (NOAA). (2017b). Land & Ocean Temperature Percentiles Jan-Dec. 2016. Asheville, NC: NOAA's National Centers for Environmental Information. Retrieved from https://www.ncdc.noaa.gov/sotc/global/201612

National Oceanic and Atmospheric Administration (NOAA). (2017c). Technical report No S CO-OPS 083: Global and Regional Sea Level Rise Scenarios for the United States. Silver Spring, MD: U.S. Department of Commerce. Retrieved from https://tidesandcurrents.noaa.gov/publications/techrpt83_Global_and_Regional_SLR_Scenarios_for_the_US_final.pdf

National Snow and Ice Data Center. (2017). Thermodynamics of Albedo. Boulder, CO: University of Colorado. Retrieved from https://nsidc.org/cryosphere/seaice/processes/albedo.html

NationMaster. (2003-2017). Countries' ranking—Economy, Health. Woolwich, Australia: NationMaster.

Nerem, R. S., Chambers, D., Choe, C., & Mitchum, G. T. (2010). Estimating Mean Sea Level Change from the TOPEX and Jason Altimeter Missions. *Marine Geodesy*, *33*(Suppl. 1), 435–446. doi:10.1080/01490419.2010.491031

Oelrichs, C. (2015). Observations on the Borneo and Sumatra fires 2015. Washington, D.C.: Global Forest Watch. Retrieved from http://www.globalforestwatch.org/stories/273

Omar, F. (2015). Unusual heat wave hits Kuwait, Middle East. *Kuwait Times*. Retrieved from http://news.kuwaittimes.net/website/unusual-heat-wave-hits-kuwait-middle-east/

Oskin, B. (2013, February 8). Ten Worst Blizzards. *LiveScience*. Retrieved from http://www.livescience.com/31880-countdown-10-worst-blizzards.html

Pandey, A. (2016, November 28). Pine Island Glacier in West Antarctica is melting from inside out. *International Business Times*. Retrieved from http://www.ibtimes.com/pine-island-glacier-west-antarctica-melting-inside-out-2451579

Richter-Menge, J., Overland, J. E., & Mathis, J. (2016). Arctic report card of the National Oceanographic and Atmospheric Administration (NOAA): executive summary. Washington, DC: U.S. Department of Commerce. Retrieved from http://www.arctic.noaa.gov/Report-Card/Report-Card-2016/ArtMID/5022/ArticleID/270/Executive-Summary

Rocha, V. (2015, March 14). Heat wave sets new temperature records across California. *Los Angeles Times*. Retrieved from http://www.latimes.com/local/lanow/la-me-ln-recordbreaking-heat-wave-broils-southern-california-20150314-story.html

Sall, C. (2013). Climate Trends and Impacts in China. Washington, D.C.: World Bank Group. Retrieved from https://openknowledge.worldbank.org/bitstream/handle/10986/17558/850550WP0P130400Impacts0in0China0EN.pdf?sequence=1&isAllowed=y

Scripps Institution of Oceanography, NOAA Earth System Research Laboratory. (2017). Trends in Atmospheric Carbon Dioxide: Full Mauna Loa CO_2 Record. Retrieved from https://www.esrl.noaa.gov/gmd/ccgg/trends/full.html

Semenza, J. C., & Menne, B. (2009). Climate change and infectious diseases in Europe. *Lancet*, *9*(6), 365–375. doi:10.1016/S1473-3099(09)70104-5 PMID:19467476

Spector, D. (2013, March 31). If you think China's air is bad, you should see its water. *Business Insider*. Retrieved from http://www.businessinsider.com/china-water-pollution-2013-3

Surtees, L. (2003). Bell, Alexander Graham. In *Dictionary of Canadian Biography* (Vol. 15). University of Toronto/Université Laval. Retrieved from http://www.biographi.ca/en/bio/bell_alexander_graham_15E.html

Sweet, W. V., Kopp, R. E., Weaver, C. P., Obeysekera, J., Horton, R. M., Thieler, R., & Zervas, C. (2017). *Global and regional sea level rise scenarios for the United States. NOAA Technical Report NOS CO-OPS 083*. Silver Spring, MD: NOAA.

Theuerkauf, E. J., Rodriguez, A. B., Fegley, S. R., & Leuttich, R. A. Jr. (2014). Sea level anomalies exacerbate beach erosion. *Geophysical Research Letters*, *41*(14), 5139–5147. doi:10.1002/2014GL060544

Trtanj, J., Jantarasami, L., Brunkard, J., Collier, T., Jacobs, J., . . . Thurston, J. (2016). Climate impacts on water-related illness. In The Impacts of Climate Change on Human Health in the United States: A Scientific Assessment. Washington, D.C.: U.S. Global Change Research Program. Retrieved from https://health2016.globalchange.gov/

Uccellini, L. W. (2016). Evolving Climate in the NWS' Weather-Ready Nation. National Climate Services Meeting. National Weather Service. Retrieved from http://www.nws.noaa.gov/om/csd/content/nationalClimateMeeting/presentations/Uccellini_Monday5.9.pdf

Union of Concerned Scientists. (2017). Confronting the realities of climate change: the consequences of global warming are already here. Retrieved from http://www.ucsusa.org/global_warming#.WJF3MIWcGhc

United Nations. (2015, November 4). Paris Climate Agreement to enter into force. Retrieved from http://www.un.org/sustainabledevelopment/blog/2016/10/paris-climate-agreement-to-enter-into-force-on-4-november/

United Nations. (2016). Marrakech Partnership for Global Action. New York: United Nations. Retrieved from http://unfccc.int/files/paris_agreement/application/pdf/marrakech_partnership_for_global_climate_action.pdf

United Nations. (2017). Sustainable Development Goals: 17 goals to transform our world. Retrieved from http://www.un.org/sustainabledevelopment/hunger/

U.S. Environmental Protection Agency. (2012). *Climate Change Indicators in the United States* (2nd ed.). Washington, D.C.: US EPA.

U.S. Environmental Protection Agency. (2016). *Inventory of U.S. Greenhouse Gas Emissions and Sinks: 1990-2014*. Washington, D.C.: U.S. Environmental Protection Agency.

U.S. Environmental Protection Agency. (2017). Can mold cause health problems? U.S. Environmental Protection Agency: Washington, D.C. Retrieved from https://iaq.zendesk.com/hc/en-us/articles/211432938-Can-mold-cause-health-problems-

USGCRP. (2016). *Impact of Climate Change on Human Health in the United States: A Scientific Assessment* (A. J. Crimmins et al., Eds.). Washington, D.C.: U.S. Global Change Research Program.

Voice of America. (2016). Hurricane Matthew weakens over U.S. after killing nearly 1,000 in Haiti. Washington, D.C.: VOA. Retrieved from http://www.voanews.com/a/weakened-hurricane-matthew-hits-georgia-carolinas/3542234.html

Wallace, S. (2017). Last of the Amazon. Washington, D.C.: National Geographic. Retrieved from http://environment.nationalgeographic.com/environment/habitats/last-of-amazon/#page=1

Walsh, J., Wuebbles, D., Hayhoe, K., Kossin, J., Kunkel, K., & Somerville, R. et al. (2014). Our Changing Climate. In J. M. Melillo, T. C. Richmond, & G. W. Yohe (Eds.), *Climate Change Impacts in the United States: The Third National Climate Assessment* (pp. 19–67). U.S. Global Change Research Program.

Water, U. N. (2014). The United Nations Inter-Agency Mechanism on All Freshwater-Related Issues, Including Sanitation. Water quality. New York: United Nations. Retrieved from http://www.unwater.org/statistics/statistics-detail/en/c/260727/

US White House. (2014). President's State, Local, and Tribal Leaders Task Force on Climate Preparedness and Resilience: Recommendations to the President. Retrieved from https://www.whitehouse.gov/sites/default/files/docs/task_force_report_0.pdf

World Bank. (2016). Population. Retrieved from http://data.worldbank.org/indicator/SP.POP.TOTL?locations=1W

World Health Organization. (2015). *WHO Estimates of the Global Burden of Foodborne Diseases. Executive Summary*. Geneva, Switzerland: WHO.

World Health Organization. (2017a). Malnutrition. Retrieved from http://www.who.int/maternal_child_adolescent/topics/child/malnutrition/en/

World Health Organization. (2017b). The Health and Environment Linkages Initiative (HELI). Vector-borne diseases. Retrieved from http://www.who.int/heli/risks/vectors/vector/en/

World Meteorological Organization. (2017). WMO confirms 2016 as the hottest year on record. Geneva, Switzerland: WMO. Retrieved from https://public.wmo.int/en/media/press-release/wmo-confirms-2016-hottest-year-record-about-11%C2%B0c-above-pre-industrial-era

Xinhua. (2010, March 30). Weather Disasters Cause RMB 200–300 Billion in Economic Damages Each Year in China. Retrieved from http://news.xinhuanet.com/video/2010-03/30/content_13270583.htm

Xinhua. (2012, February 2). Direct Economic Losses from Drought and Flooding in China Reach RMB 232.9 Billion in 2011. Retrieved from http://news.xinhuanet.com/society/2012-02/02/c_111482077.htm

KEY TERMS AND DEFINITIONS

Accumulated Cyclone Energy (ACE): The activity and destructive potential of individual tropical cyclones and entire tropical cyclone seasons (the square of the wind speed every 6 hours, scaled by a factor of 10,000 for usability).

Albedo: A non-dimensional unitless measure of how well a surface reflect solar energy.

Allergen: A substance capable of inducing an allergic reaction.

Arthropods: Invertebrates with an external skeleton, or exoskeleton, such as insects, spiders and other arachnids, and crustaceans.

Climate: The average weather conditions that persist for decades or longer.

Climate Change: Changes in features of the climate system, including average temperature, amount of precipitation, and frequency of severe weather events.

Greenhouse Gas (GHG): Gases that trap heat in the atmosphere, such as carbon dioxide, methane, and nitrous oxide, are called greenhouse gases (GHG).

Heat Wave: A period of abnormally and uncomfortably hot and unusually humid weather, typically lasting two or more days.

Incidence: A measure of the frequency with which an event, such as a new case of illness, occurs in a population over a period of time.

El Nino: A warming of the ocean surface, or above-average sea surface temperatures (SST), in the central and eastern tropical Pacific Ocean.

Morbidity: A disease or condition that reduces health and the quality of life. The morbidity rate is a measure of the frequency of disease among a defined population during a specified time period.

Mortality: Death as a health outcome; the mortality rate is the number of deaths in a defined population during a specified time period.

Premature (Early) Mortality or Death: Deaths that occur earlier than a specified age, often the average life expectancy at birth.

Prevalence: A measure of the number or proportion of people with a specific disease or condition at a specific point in time.

Surveillance: The collection, analysis, interpretation, and dissemination of health data.

Vector: An insect, tick, or other organism that transmits a disease from a plant or animal to an animal by biting it.

Weather: The state of the atmosphere at any given time and place.

Chapter 2
Impacts on Public Health

ABSTRACT

Many serious adverse public health impacts of climate change are already being felt around the globe, including record-breaking heat waves, severe air pollution, widespread water contamination that has brought a resurgence of cholera and has compromised clean drinking water and sanitation for more than one billion people worldwide, food scarcity and undernutrition from droughts and desertification, pandemics of vector-borne diseases, and increasingly frequent and severe natural hazards such as flooding, hurricanes, and earthquakes. Centralized, well-organized emergency preparedness planning is needed at the national, regional, and municipal levels to enable safe and efficient evacuations, and to minimize injuries and fatalities. In addition, effective planning to address the public health impacts of climate change is contingent on poverty reduction, and adequate access to education and healthcare for all. This chapter addresses the major public health impacts of global warming and the use of technologies in adapting to them.

INTRODUCTION

This chapter focuses on the public health impacts of global warming. Recommendations are made for adaptation strategies that use innovative technologies to prevent illness, injury, and loss of life and property from climate-change related events.

The public health impacts are discussed in order of the following seven categories (USGCRP, 2016):

DOI: 10.4018/978-1-5225-3414-3.ch002

- Temperature-related illness and death;
- Air pollution—indoor and outdoor;
- Food safety, nutrition, and availability;
- Water-borne diseases and sanitation;
- Vector-borne diseases;
- Natural hazards, and
- Mental health and well-being.

The public health impacts of climate change disproportionately affect vulnerable populations--lower-income individuals, communities of color, individuals with asthma and other respiratory diseases, pregnant women, young children, the elderly, and people with disabilities--who are most at risk of serious health consequences from the hazards posed by climate change (see Chapter 3).

TEMPERATURE-RELATED ILLNESS AND DEATH

One potentially life-threatening impact is an increase in the frequency and severity of heat waves. The elderly are particularly vulnerable to heat-related illness and fatalities during heat waves (Astrom et al., 2011). Young children also are at increased risk of heat-related morbidity (Xu et al., 2014). In 2003, a severe heat wave in Europe resulted in 15,000 deaths in France alone (CRED, 2015). Following this disastrous heat wave, several European countries developed heat wave early warning systems or national heat plans aimed at mitigating human health consequences of heat waves (Hagens & van Bruggen, 2015; Lowe et al., 2011).

In July, 1995, Chicago experienced an extreme heat wave that took the lives of about 739 people (Klinenberg, 2002). It was the deadliest weather event in Chicago history. The heat peaked on July 13, when O'Hare Airport showed a temperature of 104 degrees Fahrenheit, but the extreme humidity resulted in a heat index of 119 degrees Fahrenheit; at Midway Airport, on the south side of Chicago, the temperature was 103 degrees Fahrenheit but the heat index was a life-threatening 125 degrees Fahrenheit. The deadly effects of the heat wave were exacerbated by intense sunshine and trapped pollutants, taking the highest toll on poor city residents who did not have air conditioners or could not afford to turn on the air conditioners (National Centers for Environmental Information, 2016b). In addition, heat in inner city neighborhoods is intensified by the "heat island effect," whereby heat builds up in asphalt and buildings (US EPA, 2017a). Eric Klinenberg's book, *Heat Wave: A Social Autopsy of Disaster in Chicago* (2015), describes how unprepared Chicago was for the severity and length of the heat wave: meteorologists had warned Chicagoans that there would be a two-day heat wave. It lasted a week! By the time it was over, streets had buckled, power grids had failed, leaving residents without electricity for up to two days, and an estimated 739 people were dead, more than twice the number that died in the Great Chicago Fire of 1871 (Klinenberg, 2015).

More health education is needed about the dangers of heat waves, including how to stay cool and how to recognize signs of heat cramps, heat exhaustion, and heat stroke. Heat cramps are muscle spasms in the legs or abdomen that result from excessive water and salt loss, such as might occur among athletes during football practice outdoors on a very hot and humid day. Heat exhaustion is characterized by feelings of faintness, dizziness, and heavy sweating, whereas heat stroke is a life-threatening condition characterized by high fever, collapse, even convulsions or coma (American Red Cross, 2017). Heat exhaustion can progress to heat stroke without treatment, especially for the elderly, young children, and individuals with medical conditions. In addition, air conditioning needs to be provided for lower-income individuals who are living without it. After the Chicago heat wave of 1995, a heat wave emergency plan was put in place that provided warnings of impending heat waves and that provided free transportation to "cooling centers" for residents without air conditioning. In addition, community networks were set up to call upon isolated vulnerable elderly residents, to foster social cohesiveness. The heat wave emergency plan and social networking proved to be effective in mitigating the effects of a subsequent heat wave in 1999. (For more about the protective factors associated with social cohesiveness, see Chapter 7.)

Preparation for heat waves will minimize the risk of health problems. The first step is to stay up to date on Heat Index forecasts. In the United States, the National Weather Service's Heat Index (HI) (Figure 1) (NOAA, 2016) measures the "real feel" temperature, which combines heat and humidity. Young children, pregnant women, people who work outdoors, and older adults are vulnerable to heat exhaustion and heat stroke and should stay indoors in air-conditioning when the HI rises above 91. To remain hydrated and keep body temperature from rising, drink plenty of water (not sugary drinks), wear comfortable light-colored clothing, and remain in the shade if you do go outside. Lower SES populations who do not have air conditioning--especially young children and seniors--as well as outdoor workers are at risk of severe heat-related health problems including death. The HI is posted daily

Figure 1. National Weather Service Heat Index
Source: NOAA (2016).

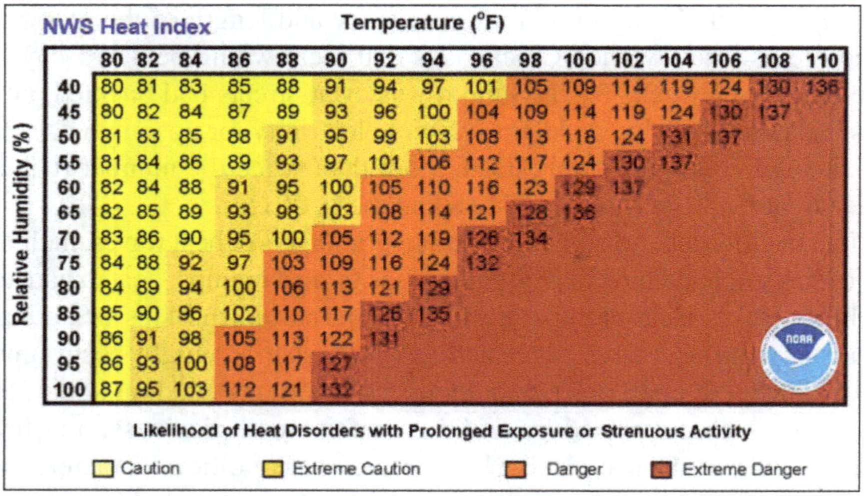

NWS Heat Index

Relative Humidity (%) \ Temperature (°F)	80	82	84	86	88	90	92	94	96	98	100	102	104	106	108	110
40	80	81	83	85	88	91	94	97	101	105	109	114	119	124	130	136
45	80	82	84	87	89	93	96	100	104	109	114	119	124	130	137	
50	81	83	85	88	91	95	99	103	108	113	118	124	131	137		
55	81	84	86	89	93	97	101	106	112	117	124	130	137			
60	82	84	88	91	95	100	105	110	116	123	129	137				
65	82	85	89	93	98	103	108	114	121	128	136					
70	83	86	90	95	100	105	112	119	126	134						
75	84	88	92	97	103	109	116	124	132							
80	84	89	94	100	106	113	121	129								
85	85	90	96	102	110	117	126	135								
90	86	91	98	105	113	122	131									
95	86	93	100	108	117	127										
100	87	95	103	112	121	132										

at http://www.nws.noaa.gov/om/heat/index.shtml or Google Air Now AQI. For those who do not have Internet access, the television weather forecasts issue heat warnings.

In addition, it is vital for municipalities to have in place a heat wave emergency plan, and to have resources available (e.g., transportation, EMT, cooling centers, ambulances) to back up the plan. Residents should be signed up for a CodeRed emergency response system, so that they can receive warnings via television, Internet, and telephone. For some vulnerable lower-income individuals without any household technology, door to door warnings and checkups are needed. With life-threatening temperatures, the only way to remain safe is to stay inside an air-conditioned building and drink plenty of water. Children, the elderly, sick and disabled individuals, pregnant women, and outdoor workers are particularly vulnerable to the health impacts of heat waves.

AIR POLLUTION

The United Nations Environmental Programme (UNEP) called air pollution "the world's worst environmental health risk" (UNEP, 2014). In its 2016 update, WHO (2017) further stated:

More than 80% of people living in urban areas that monitor air pollution are exposed to air quality levels that exceed the World Health Organization (WHO) limits. While all regions of the world are affected, populations in low-income cities are the most impacted. According to the latest urban air quality database, 98% of cities in low- and middle income countries with more than 100 000 inhabitants do not meet WHO air quality guidelines. However, in high-income countries, that percentage decreases to 56%.... As urban air quality declines, the risk of stroke, heart disease, lung cancer, and chronic and acute respiratory diseases, including asthma, increases for the people who live in them.

UNEP points out the health hazards of the indoor pollutants emitted by inefficient cookstoves that use charcoal and biomass for heating and cooking indoors; this energy source is widely used in developing nations in Asia and Africa (see *Indoor air* below).

Outdoor Air

Outdoor air quality is degraded by ground-level ozone and particulate matter (PM) pollution. Rising temperatures associated with climate change increase the concentration of ground-level ozone. When ozone remains in the upper atmosphere (the stratosphere), it shields us from ultraviolet radiation. However, at ground level, ozone creates smog, which can cause breathing problems by damaging lung tissue, reducing lung function, and inflaming airways (Carter et al., 2014). Particulate matter pollution (also called particle pollution) refers to a mixture of solid particles and liquid droplets found in the air; some, such as soot, are large enough to be visible to the naked eye. PM_{10} are inhalable particles with diameters of 10 micrometers or smaller, and $PM_{2.5}$ are fine inhalable particles with diameters of 2.5 micrometers or smaller (U.S. EPA, 2017b). The chemical makeup varies according to emissions in the area; large, coal-fired power plants emit sulfur dioxide particles, while motor vehicle exhaust emits nitrate particles (American Lung Association, 2016).

The World Health Organization (2016a) noted that in 2014, 92 percent of the world's population was living in places where the WHO air quality guidelines levels (2006) were not met. Outdoor air pollution was estimated to cause 3 million premature deaths worldwide in 2012. Policies and investments supporting cleaner transport, energy-efficient housing, power generation, industry and better municipal waste management would reduce key sources of urban outdoor air pollution. Reducing outdoor emissions from household coal and biomass energy systems, agricultural waste incineration, forest fires and certain agro-forestry activities (e.g. charcoal production) would reduce key rural and peri-urban air pollution sources in developing regions (WHO, 2016a).

In the United States, nearly half of all Americans—an estimated 150 million—live in regions that do not meet Federal air quality standards. Passenger vehicles and trucks are the main sources of this pollution, which includes ozone and particulate matter pollution (Union of Concerned Scientists, 2017).

UNEP (2014) notes that outdoor air quality is deteriorating rapidly in major cities in low and middle income countries (LMICs). The WHO (2006) guidelines for particle pollution are 20 micrograms per cubic meter for PM_{10} and 10 micrograms per cubic meter for $PM_{2.5}$. Air pollution levels in LMICs' major cities sometimes far exceed these levels. For example, in Kathmandu, Nepal, $PM_{2.5}$ levels of more than 500 micrograms per cubic meter have been measured.

According to UNEP (2014), more than 3.5 million people die each year as a result of exposure to outdoor air pollution. From 2005 to 2010, deaths from exposure to outdoor air pollution rose 4 percent worldwide, and 5 percent and 12 percent in China and India, respectively (UNEP, 2014). The costs to society of outdoor air pollution are enormous. The Organisation for Economic Co-operation and Development (OECD) estimated that the economic burden of outdoor air pollution (in US dollars) in 2010 was $1.4 trillion in China and $0.5 trillion in India. Including the world's most developed economies plus India and China, the OECD estimated total costs at $3.5 trillion per year in lives lost and ill health due to outdoor air pollution (OECD, 2016).

Symptoms of exposure to outdoor air pollution include shortness of breath, wheezing and coughing, susceptibility to infections, asthma attack, cardiovascular damage, swollen, red lung tissue, lung cancer, and premature death; in addition, there may be developmental damage to children and reproductive damage to fetuses in pregnant women (American Lung Association, 2016). Children are at greater risk of adverse health consequences from outdoor air pollution than are adults because they breathe in more oxygen per pound and tend to spend more time outdoors. In addition, people with asthma or other respiratory diseases, pregnant women and infants, the elderly, low-income individuals, communities of color (especially African Americans in the US), and outdoor workers are considered to be at higher risk of health hazards from outdoor air pollution. Individuals with asthma are at risk of worsened symptoms and asthma attacks (Carter et al., 2014). (For more details on climate change and vulnerable populations, see Chapter 3.)

The U.S. Environmental Protection Agency's (EPA) Air Quality Index (AQI) measures the level of outdoor air pollution. When the AQI is in the unhealthy range, people should stay indoors in air-conditioning as much as possible; when they do go outside, they should avoid exertion and heavily trafficked roadways and consider wearing a mask. The local AQI is posted daily at https://www.airnow.gov/.

Outdoor workers' exposure to ozone and particle pollution is much more extensive than is that of indoor workers. In Galveston, Texas, after it was discovered that unsafe ozone levels caused airway obstruction in healthy young lifeguards, the city implemented an air quality warning flag system on Galveston's beaches to alert residents when ozone levels were unhealthy (American Lung Association, 2016).

The Air Quality Flag Program (AQFP) was developed by the Environmental Protection Agency for use by municipalities, schools, health departments, fire stations, businesses, museums, and other organizations to alert residents when the outdoor air quality is unhealthy (U.S. Environmental Protection Agency, 2016). Warning flags are particularly helpful for at-risk populations such as low-income communities of color where access to online AQI may not be readily available, and for pregnant women, children, senior citizens, and those with asthma and other lung diseases. The AQFP is color-coded into five categories (U.S. EPA, 2016) (see Figure 2).

A green flag indicates the air quality is healthy for all individuals. A yellow flag indicates that there is moderate air pollution, not enough to affect most individuals; however, doctors may advise that frail and extremely sensitive or ill individuals stay indoors. An orange flag indicates that the air quality is unhealthy for sensitive groups such as those with asthma or other lung disease, pregnant women, children, senior citizens, low-income communities of color, and outdoor workers. A red flag means that the air quality is unhealthy for all individuals, while a purple flag indicates hazardous air quality. The EPA's

Figure 2. Air Quality Warning Flags
Source: U.S. Environmental Protection Agency (2016).

Air Quality Flag Program (U.S. EPA, 2016) is a technological innovation that can be used to inform the public about outdoor air quality and to warn vulnerable individuals when they should stay inside. A free Smartphone app can be downloaded from the American Lung Association (2017a) website.

The AQFP proved to be invaluable in summer 2016 in alerting residents of Big Sur and Carmel in Monterey County, California about the wildfire smoke impacts when the Soberanes wildfire burned more than 132,000 acres over a three-month period (US Environmental Protection Agency, 2016). The EPA would like to see the AQFP in use at all schools; currently, less than 1 percent of schools participate. Implementing the AQFP at schools would help students with asthma to avoid a severe exacerbation and visit to the Emergency Department by alerting them when the AQFP is unhealthy for them. Students without asthma could be kept inside at recess instead of playing outdoors (US Environmental Protection Agency, 2016).

Indoor Air

Indoor air quality (also called "indoor environmental quality") is determined by factors such as temperature, humidity, and ventilation, as well as exposure to chemicals or pollutants; it can affect a person's health, comfort, and ability to work (Occupational Safety and Health Administration, 2016). Indoor air quality is degraded by airborne allergens, mold, mildew, and volatile organic compounds (VOCs) from tobacco smoke, paint and paint remover, cleaning products, pesticides, air fresheners, personal care products such as hairspray, hobby products such as glue, wood burning stoves, office equipment such as printers and copiers, pressed wood furniture, carpeting, fuel oil and gasoline, and car exhaust (from attached garage).

In developing countries, indoor air quality is degraded by the burning of coal and biomass, used by about 3 billion people worldwide to cook and heat their homes (UNEP, 2014). In 2012, 4.3 million deaths were attributed to exposure to household air pollution, almost all in LMICs. Thirty-four percent of the deaths were from Stroke, 26 percent from Ischemic Heart Disease, 22 percent from Chronic Obstructive Pulmonary Disease (COPD), 12 percent from Acute Lower Respiratory Disease, and six percent from Lung Cancer. Forty-six percent of the deaths were men aged 25 and older, 41 percent were women 25 and older, and 13 percent were children under age 5 (WHO, 2014a). The Southeast Asian and Western Pacific regions had by far the highest burden, with 1.69 and 1.62 million deaths, respectively. Sub-Saharan Africa was third highest, with almost 600,000 deaths (WHO, 2014a). The World Bank (2011) found that providing incentives to households to switch from wood-based energy sources to alternate energy sources did not yield the desired results. So, an effort has been made to modernize the coal and wood-based biomass burning in sub-Saharan Africa, where is it the dominant household energy source and also a major source of employment. Improved cookstoves have been developed that are fuel efficient and emit fewer pollutants. Improved, more efficient kilns have been developed for the conversion of wood into charcoal. Adoption of innovations could be incentivized and promoted using the Diffusion of Innovations theoretical framework (see Chapter 5).

A technical innovation that potentially could provide a solution to the biomass emissions problem is the production of sustainable charcoal, which would be carbon neutral since the emitted carbon would be sequestered by trees. In carbon sequestration, trees absorb carbon dioxide, release the oxygen, and store the carbon. However, the process of sustainable harvesting of wood and conversion of wood into charcoal is plagued by many problems, including illegal harvesting and other corrupt practices (World Bank, 2011). The World Bank concluded that developing a sustainable charcoal production would require a reform and modernization of the fiscal and governance framework, including formalizing charcoal trade, linking fiscal policies to sustainable harvests and sustainable charcoal, enhancing and modernizing law enforcement, investing in infrastructure (trading sites), promoting the use of information technology, piloting sustainable charcoal certification, and fostering analytical work and knowledge exchange (World Bank, 2011).

Many households have unhealthy indoor air that can cause breathing or other health problems. If you notice that your symptoms get better outside of your home, there may be sources of air pollution in your home. Use the following checklist from the American Lung Association (2017b) to help pinpoint the potential source(s) of the problem:

- Is anyone smoking tobacco indoors? No one should smoke indoors.
- Can you see or smell mold or mildew?
- Is the humidity regularly above 50 percent?
- Are there leaks or standing water anywhere—kitchen, basement, attic?
- Are all fuel-burning appliances (gas stoves, water heaters, fireplaces) fully vented to the outdoors?
- Is there an attached garage or basement where cars, lawnmowers or motorcycles are stored?

- Are household chemicals, paints or solvents stored indoors or in an attached garage or basement?
- Have you recently remodeled, added new furniture or carpeting, or painted?
- Do you use odor-masking chemicals or "air-freshening" devices?
- Has kitchen or food garbage been covered and removed?
- Have you used pesticides recently?
- Have you tested your home for radon? Although radon doesn't cause noticeable, physical symptoms, you should test your home for this dangerous substance.

If you suspect that your indoor air may be polluted, check the temperature, humidity, and air flow. Change filters on air conditioning and furnaces and have the systems serviced regularly. Test your home for radon. Indoor air quality monitors can be purchased to monitor air quality inside homes. One such monitor, Foobot air quality monitor, connects to a phone app that enables monitoring pollutants and allergens in indoor air; Foobot is fully integrated with IFTTT, enabling the user to set up an if, then action—e.g., if the pollution reaches a certain level, turn on air purifier (Cawley, 2016). More economical models such as Legato Eve Room ($79.99) are available; Legato Eve Room comes with Apple Homekit technology, for connection to Iphone or Ipad.

According to the U.S. Department of Labor, in the workplace, most common building problems associated with indoor air quality are due to inadequate ventilation, lack of fresh outdoor air, or contaminated air being brought into the building; poor maintenance of ventilation, heating, and air conditioning systems; dampness and moisture damage due to leaks, flooding, or high humidity; and occupant activities such as construction and remodeling (Occupational Safety and Health Administration, 2016). If an employee suspects that the indoor air quality in the workplace is degraded, he or she should first talk to the employer, who may check temperature, humidity, and ventilation. If talking to the employer does not correct the problem, the employee may contact the U.S. Labor Department's Occupational Safety and Health Administration (OSHA) at their toll-free number (800-321-OSHA (6742) or TTY 1-877-889-5627) or request a Health Hazard Evaluation (HHE) from the National Institute for Occupational Safety and Health (NIOSH). NIOSH provides free investigations of workplace health hazards in response to requests from employers, employees and their representatives, and federal agencies; the NIOSH website provides forms and information about requesting a Health Hazards Evaluation (NIOSH, 2016).

It is vital that schools perform an Indoor Air Quality (IAQ) assessment to make sure that the air ventilation meets the standards of the American Society of Heating, Refrigerating, and Air-Conditioning Engineers (ASHRAE, 2017). Studies have shown that good IAQ reduces student absenteeism and improves test scores (US EPA, 2014). Healthy learners are better learners (Basch, 2011)! While some may protest that making a school building's indoor air safe is too costly to undertake, recent case studies have proved that becoming energy efficient and healthy actually saves the school district a bundle! For example, a recent retrofit of an entire school district's HVAC systems recouped its costs in one year and has gone on to save the district $11.2 million since 2008. All 31 schools in this district, in Vancouver, Washington, are enrolled in the Green Schools Program of Washington State and met the ENERGY STAR certification for Healthy School Buildings (ENERGY STAR, 2017).

FOOD SAFETY, NUTRITION, AND AVAILABILITY

Food security—ongoing access to a safe, nutritious, adequate food supply-- is impacted by rising temperatures, droughts, floods, and extreme weather events associated with climate change. The food system interacts with our physical and biological environments as food moves from "farm to table." Rising temperatures and flooding can increase pathogen load, and increase vectors and pests (and, therefore, pesticide use). Warmer temperatures increase the risks of food spoilage, and extreme climate events disrupt food distribution (Ziska et al., 2016). Rising CO_2 exposure can decrease nutritional quality of foods, by lowering the amount of protein and essential minerals in a number of widely consumed crops, including wheat, rice, and potatoes (Loladze, 2014).

In July 2016, U.S. President Obama signed the Global Food Security Act of 2016, which allocated more than $7 billion to initiatives that focus on agriculture, small-scale food producers, and the nutrition of women and children worldwide. The new law reinforced the 2009 U.S. initiative, Feed the Future. In 2015, Feed the Future helped more than 9 million small-scale farmers and rural families adopt practices that improved agricultural productivity, boosting their incomes by more than $800 million (Crescente, 2016).

Eradicating hunger and malnutrition is one of the most urgent global problems, especially in developing nations. Adequate nutrition is necessary for healthy physical and cognitive development. Malnourished children have stunted growth and their ability to access learning is impaired. In 2015, 193 world leaders adopted the 17 Global Goals for Sustainable Development to end extreme poverty, hunger, and inequality, and to mitigate climate change through sustainable development by the target year 2030. Goal # 2 is Zero Hunger—pledges to end hunger, achieve food security, improve nutrition, and promote sustainable agriculture (United Nations, 2017). One in nine people in the world today (795 million) is undernourished. The vast majority of the hungry live in developing countries, where 12.9 percent of the population is undernourished. Asia accounts for two-thirds of the total undernourished people. In sub-Saharan Africa, projections for the 2014-2016 period indicate that almost 23 percent of the population is undernourished. Poor nutrition causes 45 percent of deaths of children under five – 3.1 million children each year. A quarter of the world's children suffer from stunted growth due to inadequate nutrition (United Nations, 2017).

A major initiative to address the Zero Hunger Goal for Sustainable Development is the World Food Programme's Food Assistance for Assets (FFA) program, which addresses immediate food needs with cash, vouchers, or food transfers while tackling the longer-term goal of food security and resilience by building or rehabilitating assets such as crops, forests, barren lands, and irrigation systems. Since 2013, FFA programs have helped 10 to 15 million people per year in 50 countries to bring degraded lands back into productive use. FFA's projects are sustainable, with built-in climate resilience to help communities survive extreme weather events such as hurricanes without losing their means of food production. FFA's programs reduce erosion and desertification and improve soil conditions; they strengthen and diversify income and livelihoods, promote gender equality, and empower communities to work together to end hunger (FFA, 2017).

In Chapter 7 of *The Impacts of Climate Change on Human Health in the United States: A Scientific Assessment* (Ziska et al., 2016), the following key findings are noted:

- Increased exposure of food to pathogens and toxins, increasing foodborne illnesses;
- Increased human exposure to chemical contaminants in food from mercury in seafood (caused by elevated sea surface temperatures), contaminants in the food chain (from extreme weather events), and increased levels of pesticides and veterinary drugs in food chain (due to increase in pests, vectors, and microbes);
- Decreased nutritional value of important food crops such as wheat and rice due to rising levels of CO_2;

- Disruptions of food distribution due to increase in extreme weather events.

In 2015, the World Health Organization published the WHO Initiative to Estimate the Global Burden of Foodborne Diseases, which estimated global foodborne disease incidence, mortality, and disease burden. The report analyzed 31 foodborne hazards and found that low-income regions and young children were disproportionately affected, with 40 percent of the foodborne diseases among children under age five. The 31 hazards caused 600 million foodborne illnesses and 420,000 deaths worldwide in 2010. There was considerable variation in the burden of disease, with Africa being the most afflicted, followed by Southeast Asia and the Eastern Mediterranean (WHO, 2015).

In contrast to the developing world, the United States has one of the safest food supplies. Yet, *Campylobacter* bacteria was found on 47 percent of raw chicken samples tested in grocery stores in 2011 (CDC, 2014). *Campylobacter* and *Salmonella* were the two most common bacteria responsible for foodborne diseases. Chickens infected with *Campylobacter* spread the bacteria through a common water source or infected feces. *Salmonella* comes from animal feces, and most commonly is found in raw or undercooked eggs, but is also found in chicken and produce. The Centers for Disease Control and Prevention (CDC) estimated that there are 48 million cases of foodborne illnesses per year, with approximately 3,000 deaths (CDC, 2015). There are seasonal peaks for the foodborne illnesses: infections from *Campylobacter, Salmonella,* and *Escherichia coli (E. coli)* peak in the summertime, while infections from *Noroviruses* peak in the wintertime (Lal et al., 2012; Ahmed et al., 2013).

There are four steps to remember when preparing and eating food to prevent food poisoning:

- Clean—wash your hands for 20 seconds with plain soap and running water; wash surfaces and utensils after each use; wash produce but not meat, poultry or eggs.
- Separate—use separate cutting boards and plates for produce and for meat, poultry, seafood, and eggs; keep meat, poultry, seafood, and eggs separate from all other foods at the grocery store and in your refrigerator.
- Cook—use a food thermometer and keep food hot after cooking (at 140 degrees Fahrenheit or above); microwave food thoroughly (to 165 degrees Fahrenheit).

- Chill—refrigerate perishable food within two hours (one hour in hot weather [temperature 90 degrees Fahrenheit or above]); never marinate or thaw foods on the counter; know when to throw out food (See Tables 1 & 2).

Videos are available on the FoodSafety.gov site that give demonstrations of safe food handling by a family who uses the four safety principles of clean, separate, cook, and chill.

WATERBORNE DISEASES AND SANITATION

The World Health Organization (Berman, 2009) called waterborne disease the "world's leading killer." Most of its 3.4 million annual fatalities are

Table 1. Safe storage times for meat, poultry, and tuna salad

Category	Food	Refrigerator (40 °F or Below)	Freezer (0 °F or Below)
Salads	Egg, chicken, ham, tuna & macaroni salads	3 to 5 days	Does not freeze well
Hot dogs	opened package	1 week	1 to 2 months
	unopened package	2 weeks	1 to 2 months
Luncheon meat	opened package or deli sliced	3 to 5 days	1 to 2 months
	unopened package	2 weeks	1 to 2 months
Bacon & Sausage	Bacon	7 days	1 month
	Sausage, raw — from chicken, turkey, pork, beef	1 to 2 days	1 to 2 months
Hamburger & Other Ground Meats	Hamburger, ground beef, turkey, veal, pork, lamb, & mixtures of them	1 to 2 days	3 to 4 months
Fresh Beef, Veal, Lamb & Pork	Steaks	3 to 5 days	6 to 12 months
	Chops	3 to 5 days	4 to 6 months
	Roasts	3 to 5 days	4 to 12 months
Fresh Poultry	Chicken or turkey, whole	1 to 2 days	1 year
	Chicken or turkey, pieces	1 to 2 days	9 months
Soups & Stews	Vegetable or meat added	3 to 4 days	2 to 3 months
Leftovers	Cooked meat or poultry	3 to 4 days	2 to 6 months
	Chicken nuggets or patties	3 to 4 days	1 to 3 months
	Pizza	3 to 4 days	1 to 2 months

Source: FoodSafety.gov (2017).

Table 2. Safe storage times for eggs and products made with eggs

Raw Eggs in Shell	3 to 5 Weeks	Do not Freeze. Instead, Beat Yolks and Whites Together; Then Freeze.
Raw egg whites	2 to 4 days	12 months
Raw egg yolks	2 to 4 days	Yolks do not freeze well.
Raw egg accidentally frozen in shell	Use immediately after thawing.	Keep frozen; then refrigerate to thaw.
Hard-cooked eggs	1 week	Do not freeze.
Egg substitutes, liquid *Unopened*	10 days	12 months
Egg substitutes, liquid *Opened*	3 days	Do not freeze.
Egg substitutes, frozen *Unopened*	After thawing, 7 days or refer to "Use-By" date.	12 months
Egg substitutes, frozen *Opened*	After thawing, 3 days or refer to "Use-By" date.	Do not freeze.
Casseroles with eggs	3 to 4 days	After baking, 2 to 3 months.
Eggnog *Commercial*	3 to 5 days	6 months
Eggnog *Homemade*	2 to 4 days	Do not freeze.
Pies Pumpkin or pecan	3 to 4 days	After baking, 1 to 2 months.
Pies Custard and chiffon	3 to 4 days	Do not freeze.
Quiche with filling	3 to 4 days	After baking, 1 to 2 months.

Source: FoodSafety.gov (2017).

young children; the vast majority of these children die from illnesses caused by organisms that thrive in water contaminated by raw sewage. Cholera, typhoid fever and hepatitis A are common diarrheal diseases caused by bacteria; dysentery and other illnesses are caused by parasites that live in water contaminated by the feces of sick individuals. People who live without running water or latrines use the same lake or stream for drinking water, bathing and defecating, contaminating the water. Water supplies are severely compromised by lack of latrines; 18 percent of the world's population defecates in the open. UNICEF (2016) reports that 1.1 billion people still do not have access to safe, clean water. Seventy percent of the world's population who lack sanitation—1.8 billion people—live in Asia (WHO/UNICEF, 2008). In sub-Saharan Africa, 180,000 children under five years of age die every year—about 500 per day—due to diarrheal diseases linked to inadequate water, sanitation, and hygiene. In the past 25 years the population of sub-Saharan Africa has nearly doubled while access to sanitation and water have increased by only 6 percentage points and 20 percentage points, respectively (UNICEF, 2015). Climate change events such as extreme weather, droughts, and heat waves have exacerbated the sanitation and water problems.

People also may catch a diarrheal disease by eating food prepared or served by a sick individual who has not washed his or her hands. All people need to have access to a nearby source of clean drinking water. When people have to walk miles to find clean water, they are not going to wash their hands very often! When people have no latrines, and defecate in the open, the feces contaminate food sources, especially after heavy rains or flooding. Feces also may be tracked into homes at other times by people or animals, spreading infectious organisms throughout the living quarters (Berman, 2009).

Water pollution in China is at crisis proportions. In March 2013, more than 2,000 dead pigs were found floating in a Shanghai river, the main water source for the city's population of 23 million (Spector, 2013). More than 80 percent of water from wells used in rural China is unfit for drinking or bathing because of contamination from industry and farming (Buckley & Piao, 2016). There are reports of "cancer villages" caused by drinking contaminated water (Spector, 2013). In addition, outbreaks of harmful toxin-producing bacteria have increased; in 2015 there was a severe blue algae outbreak in Chaohu Lake in eastern China, the country's fifth largest freshwater lake. In June 2015, a "green tide" of algae 35,000 square km across was seen in waters off China's eastern coast (Mengjie, 2015).

India is the country with the greatest number of urban-dwellers who do not have safe private toilets and who practice open defecation. The rapid urbanization without the infrastructure to support sanitation has led to overflowing urban slums where 157,191,000 people live without toilets (WaterAid, 2016). The high population density of urban areas exposes people to the pollution of others, who live in close proximity. Narrow crowded streets are not always passable by waste management trucks. In addition, rising sea levels have caused higher ground waters in coastal areas, which are contaminated by raw sewage seeping through pits or being discharged into wastewater systems. India has the highest number of urban dwellers living without toilets; China is not far behind. Other countries on the top ten list are Nigeria, Indonesia, Russian Federation, Bangladesh, DR Congo, Brazil, Ethiopia, and Pakistan (WaterAid, 2016).

Goal #6 of the 17 Global Goals for Sustainable Development is to ensure access to water and sanitation for all by 2030. Specifically, it aims to achieve universal and equitable access to safe and affordable drinking water for all, to achieve access to adequate and equitable sanitation and hygiene for all, to end open defecation, to improve water quality by reducing pollution, eliminating dumping and minimizing release of hazardous chemicals and materials, halving the proportion of untreated wastewater and substantially increasing recycling and safe reuse of water (United Nations, 2017). The WHO/UNICEF Joint Monitoring Programme for Water Supply and Sanitation (JMP) was established in 1990 to monitor and promote clean drinking water, sanitation, and hygiene (WASH). The JMP has been instrumental in 1) developing norms used to benchmark progress in access to WASH, 2) developing targets, policies, and investment programs for reducing WASH inequalities, and 3) facilitating a critical dialogue among WASH stakeholders on key trends and then developing new approaches to monitor progress. A particular concern of JMP was to improve access to WASH for girls and women for their menstrual hygiene (WHO/UNICEF JMP, 2015).

Waterborne illnesses are caused by the following agents: 1) pathogens, such as bacteria, viruses, and protozoa; 2) toxins produced by certain harmful algae; and 3) chemicals introduced into the environment by human activities. Human exposure occurs through ingestion, inhalation, or direct contact with contaminated drinking or recreational water, as well as through consumption of contaminated fish and shellfish. In the United States, approximately 97 percent of all waterborne illnesses are caused by the following eight pathogens: the enteric viruses norovirus, rotavirus, and adenovirus; the bacteria Campylobacter jejuni, E. coli O157:H7, and Salmonella enterica; and the protozoa Cryptosporidium and Giardia (Soller et al. 2010).

In Chapter 6 of *The Impacts of Climate Change on Health in the United States* (Trtanj et al., 2016), the following key findings on waterborne diseases were highlighted:

- Increased water temperatures associated with climate change will alter the seasonal windows of growth and geographic range of suitable habitats for harmful toxin-producing algae and bacteria.
- Runoff from more frequent and intense precipitation will compromise recreational water, shellfish harvesting, and drinking water through the introduction of pathogens and toxic algae blooms.
- More frequent and severe extreme weather events and storm surges will increase the risk of infrastructure failure for drinking water, wastewater, and storm water systems.

Intense, prolonged droughts associated with climate change have been causing severe water security problems in the Middle East, sub-Saharan Africa, China, and other regions around the globe. There is great promise in the technological innovations developed by Israel, whose "water revolution" has turned around a drought-stricken land that is 60 percent desert into a water management leader. Israel's water expertise is now being sought in China, the United States, and around the world. The five principles of the Israeli water revolution are as follows: 1) education and conservation, taught to schoolchildren and promoted on TV, 2) wastewater reclamation and reuse (Israel currently reuses 86 percent of its wastewater for agricultural use), 3) government policy (centralized control over water sources, which belong to the state), 4) water and agricultural technology (Israel invented drip irrigation), and 5) desalination (five plants provide more than 25 percent of the nation's water supply and more than 80 percent of household water). The success of Israel's water revolution rests on its combination of education, conservation, and the use of innovative technology (Distel, 2016). More than half of Israel's water comes from desalination and water recycling, providing a role model for other drought stricken regions, including California (Grossman, 2016). By comparison, Spain, which is second in water recycling, recycles only 17 percent of its water, and the United States, only 1 percent of its water (Kershner, 2015)!

VECTOR-BORNE DISEASES

The European Centre for Disease Prevention and Control defines vector-borne diseases as "infections transmitted by the bite of infected arthropod species, such as mosquitoes, ticks, triatomine bugs, sandflies, and blackflies" (Confalonieri et al., 2007). Infected vectors carry pathogens (viruses, bacteria, protozoa) that can be transferred from one host to another, for example, when a tick infected with Lyme disease bites a human.

Climate change, especially temperatures changes and level of precipitation, affect vector distribution and prevalence. The vector distribution is affected by a host of other factors such as habitat destruction, land use, pesticide application, and host density (Semenza & Menne, 2009).

In Chapter 5 of *The Impacts of Climate Change on Health in the United States* (Beard et al., 2016), the following key findings on vector-borne diseases were highlighted:

- Changing of geographic and seasonal distributions of vectors and vector-borne diseases due to climate change;
- Earlier tick activity and a northward range expansion due to rising temperatures caused by climate change;
- Changing of distribution, abundance, and prevalence of infection in mosquitoes due to climate change, with a resulting increase in human exposure to bites and increased risk of contracting a vector-borne disease;
- Emergence of new vector-borne pathogens due to the interaction of climate change factors with changing land-use patterns; however, risk of human disease can be reduced by adaptation that includes effective vector control and human protective measures (Beard et al., 2016).

The most common vector-borne disease in the United States is Lyme disease; the Centers for Disease Control and Prevention notes that the 39,307 cases reported by health departments in 2013 represent about a tenth of the case rate, which is estimated at 329,000 per year (Beard et al., 2016; CDC, 2015). Lyme disease is caused by the bacterium *Borrelia burgdorferi sensu stricto* and is transmitted by ticks in the nymph stage, which occurs in the summer months. It is transmitted to humans in the eastern United States primarily by the tick species *Ixodes scapularis*, known as blacklegged ticks or deer ticks, and in the far western United States by *I. pacificus*, commonly known as western blacklegged ticks (Beard et al., 2016). The migration of deer into suburban areas due to reduction of their habitat has caused increased contact between infected deer and humans and their pets. Deer wander into yards and ticks get on plants and bushes and then onto dogs or humans, particularly children, who play on the ground. Lyme disease is so prevalent in the Northeastern United States that public health campaigns have urged people to wear long sleeves and pants with socks and closed shoes when walking in wooded areas, to use repellant, and to check the body, face, and neck for ticks after walking in wooded areas. With Lyme disease, early detection is critical; late-stage (tertiary) Lyme disease can chronic arthritis, joint inflammation, memory loss, and other damage.

The most common mosquito-borne disease in the United States is the West Nile Virus (WNV), which was discovered in Uganda in 1937 but was not viewed as a public health threat until it caused epidemics in the Middle East in the 1950s. It continued to spread around the world, reaching the United States (New York City) in 1999 (Semenza & Menne, 2009). WNV can thrive in a wide variety of ecosystems and has been found in 65 different varieties of mosquito in the United States. WNV mainly affects birds. It is difficult to know the true number of human infections, because many of those infected with the West Nile Virus are asymptomatic (Semenza & Menne, 2009). There were 2,038 cases of West Nile Virus reported to the Centers for Disease Control and Prevention in 2016 (CDC, 2017a).

In 2016, the Zika virus emerged in the United States, mostly among travelers who contracted it on vacation. The Zika virus is transmitted by an infected *Aedes* species mosquito (*Ae. aegypti* and *Ae. Albopictus*). Given its link to hydrocephaly in babies born to women infected with the Zika virus during pregnancy, there has been great concern about prevention. Puerto Rico, a U.S. Territory, is a hotspot for Zika, with a preliminary report of 34,577 cases in 2016 (CDC, 2017b). Mosquitoes only need a half cup of water to breed, so mitigation measures include emptying all containers of standing water and not leaving them out where they can fill up with water during heavy rains. Recycling bins should be securely closed, as the empty bottle and cans will fill with water during storms and then could attract mosquitoes. Some mosquitoes have become resistant to insecticides.

In the 1950s-1960s, indoor spraying programs brought vector-borne diseases under control; subsequently, vector-control programs lapsed and resources dwindled, leading to a resurgence of vector-borne diseases in the past two decades (WHO, 2014b). In addition, the vector-borne diseases, traditionally regarded as a tropical problem, have been spreading globally, affecting new populations and geographic areas. This spread is accelerated by "changes in climate, ecology, land use patterns, and the rapid and increased movement of people and goods, threatening more than half the world's population" (WHO, 2014b, p. 10). Another serious concern is the increasing resistance of many vectors to insecticides, along with a dearth of qualified entomologists and vector-control experts.

The World Health Organization (Chan, 2014, p. 7) recommends using natural predators in an integrated vector management approach:

This approach uses a range of interventions, from indoor residual spraying to the use of natural insect predators, in combination and in a value-added way. Integrated management makes sense, as many vector-borne diseases overlap geographically, some vectors cause several diseases, and some interventions provide protection against several vectors.

A promising new natural mosquito repellant and insecticide is being developed by the Centers for Disease Control and Prevention (CDC, 2016) using Nootkatone, a natural ingredient found in grapefruit and Alaska yellow cedar trees. Nootkatone repels mosquitoes and ticks when applied to the skin, and it kills them when used as an insecticide. The CDC is working with a commercial partner, Evolva, to create insecticide sprays as well as products such as lotions and soaps for use as insect repellants.

Other technological advances in the arena of vector control are the development of ArboNET, a nationwide electronic surveillance system that links health departments to the CDC to track vector-borne viruses spread by mosquitoes in the United States, and TickNET, which does the same thing for tick-borne diseases. New cutting-edge lab tests are being developed to improve diagnosis of vector-borne diseases; vaccines are being developed as well. Clinicians and health facilities are updated on the prevention and treatment of vector-borne diseases. A public health campaign is a very important adaptation measure. People have to understand how vector-borne diseases spread and what they can do to reduce their risk of contracting one.

In sub-Saharan Africa, despite improvement in case rates, malaria remains the deadliest vector-borne disease, killing more than 1.2 million children a year, most of them under age five (WHO, 2016b). New strategies for controlling the spread of diseases through mosquitoes are encompassed under Integrated Vector Management (IVM), and include environmental management strategies that reduce or eliminate vector breeding grounds through 1) improved design or operation of water resources development projects and 2) the use of biological controls (e.g. bacterial larvicides and larvivorous fish) that target and kill vector larvae without generating the ecological impacts of chemical use. At times, IVM uses chemical methods of vector control, such as insecticide-treated nets (ITNs), indoor residual sprays, space spraying, and use of chemical larvicides and adulticides. ITNs have been effective in reducing rates of malaria among children (WHO, 2016b). Public health campaigns are vital to help to reduce the transmission of malaria.

The Division of Vector-Borne Diseases (DVBD) of the U.S. Centers for Disease Control and Prevention (CDC, 2016) provides guidelines and new technologies that will help with primary prevention (avoiding getting a vector-borne disease) and secondary prevention (screening tests for vector-borne diseases that detect them shortly after transmission, when treatment is most effective). DVBD's key initiatives (CDC, 2016) are as follows:

- Developing new cutting-edge laboratory tests to identify and diagnose vector-borne diseases (e.g., Zika, Chikungunya, and dengue fever);
- Developing guidelines and educating clinicians;
- Monitoring vector-borne diseases through surveillance systems, including ArboNET, which tracks vector-borne diseases, and TickNET, which supports research on identifying tick-borne diseases;
- Researching vaccines;
- Partnering with state, local, territorial, and tribal health departments, industry, and the World Health Organization, to deal with vector-borne disease outbreaks;
- Educating the public, Congress, and other stakeholders;
- Finding effective natural insect repellants (e.g., Nootkatone).

NATURAL HAZARDS (EXTREME WEATHER EVENTS)

Earthquakes, droughts, floods, hurricanes, and other natural disasters have been escalating in frequency and severity as a result of global warming. Floods and other water-related disasters account for 70 per cent of all deaths related to natural disasters (United Nations, 2017).

The Identification of Global Natural Disaster Risk Hotspots Project is a global, multihazard risk analysis focused on identifying key geographic "hotspots" where the risks of natural disasters are particularly high (Dilley et al., 2005). The frequency of natural disasters appears to be increasing; the Emergency Events Database (EM-DAT), a global disaster database maintained by the Centre for Research on the Epidemiology of Disasters (CRED) in Brussels, records more than 600 disasters globally each year. In addition to causing huge losses of life and property, natural disasters devastate the livelihoods of the poor, wiping out their development gains. The Hotspots Project aims to identify and prioritize where development investment is most needed to reduce disaster risk (Dilley et al., 2005). It employs three categories of risk: mortality risks, risks of total economic losses, and risks of economic losses as a proportion of the GDP. Analysis of data shows that 3.4 billion people are relatively highly exposed to at least one hazard, 790 million people are relatively highly exposed to at least two hazards, and 105 million people are relatively highly exposed to three or more hazards. Hazards are divided into three categories: drought, geophysical hazards such as earthquakes and volcanoes, and hydro hazards such as floods, landslides, and cyclones. (Dilley et al., 2005). The nation with the highest risk from three or more hazards was Taiwan, with 73.1 percent of its population and 73.1 percent of its area exposed to four hazards; second was Costa Rica, with 41.1 percent of its population and 36.8 percent of its area exposed to four hazards. The United States had 11.2 percent of its population and 2.6 percent of its area exposed to two hazards. In high-risk nations and regions, development projects must reduce the infrastructure's vulnerability and exposure, and build up emergency funds and institutional, policy and capacity-building measures to improve disaster risk management.

A sampling of natural disasters around the world from May to August 2016 gives an idea of the magnitude of extreme weather events occurring worldwide; heat waves are included but were not in the Hotspot analysis:

- Flooding:
  - In China's Yangtze Basin from May through August killed at least 475 people and caused $28 billion in losses, and in northeast China in July killed 289 people and caused about $4 billion in damage (Borenstein, 2016).
  - In Sudan and South Sudan in July and August killed 129 people and damaged more than 41,000 buildings (Borenstein, 2016).
  - In West Virginia and the mid-Atlantic in June killed 23 people and damaged more than 5,500 buildings (Borenstein, 2016).
- Droughts:
  - In India that caused about $5 billion in damage. (Borenstein, 2016).
  - In Western U.S., 17-year drought, the longest recorded for this region in the historical record (Lindsey, 2014).
  - In Syria, continuation of 18-year drought, likeliest its worst in 900 years (Cook et al., 2016).
  - In northern Somalia by the end of the summer, caused widespread crop failure, with an estimated 300,000 children suffering from severe malnutrition and more than 100 children dying every day (Mountain, 2016).
- Typhoons:
  - In the Phillipines, Taiwan and China in July, killed 111 people and caused at least $1.5 billion in damage.
  - In Japan, China and Korea in August, killed 77 people and damaged more than 20,000 buildings.

- Heat Waves:
  - In May, hottest temperature ever recorded in India (51 degrees Celsius, 123.8 degrees Fahrenheit) (BBC News, 2016).
  - In July in Iran and Kuwait, a record high of 51 degrees C (124 °F) in Baghdad on Thursday, July 30 (Omar, 2015).
  - In Spain, hottest September temperature recorded in Europe of 45.3 degrees Celsius (113 degrees Fahrenheit) (Borenstein, 2016).

Mitigation of the impact of extreme weather events must include a public health campaign to reach out to vulnerable populations to educate them about how to stay safe in an emergency and to sign them up for CodeRed emergency response, weatherproofing homes against storms, providing air conditioners and/or transport to cooling centers for heat waves, and building earthquake resistant structures. Elevating and grading roadways helps to mitigate flooding; new beachfront development needs to be set back to avoid future flooding from rising sea levels. Empowering communities through the development of social networks builds resilience and is key to reducing injury and deaths (see Chapter 7).

Of vital importance is the development and employment of emergency preparedness plans, at the national, state, county, and municipal level. In 2004, a tsunami early warning system could have saved tens of thousands of lives when a massive 9.1 earthquake hit Indonesia. Such a warning system had been in existence since 1949 in the south Pacific but was not used for the Indian Ocean peoples. More than 230,000 people in 14 countries died from the tsunamis that followed the earthquake. Three days before Hurricane Katrina hit the Gulf Cost in 2005, the Governors of Louisiana and Mississippi declared states of emergency. However, the Mayor of New Orleans waited until 24 hours before the hurricane hit to give evacuation orders and there was no transportation available for those without cars. As a result, tens of thousands were stranded and failed to evacuate. In both disasters, effective emergency preparedness would have enabled a timely evacuation and greatly reduced fatalities. A coordinated, centralized emergency management operation is a prerequisite for disaster risk management (Global Assessment Report, 2015). In addition, effective disaster risk management is contingent on poverty reduction, and adequate access to education and healthcare; it cannot be achieved if the loss of schools, health facilities, housing and local infrastructure through extensive disasters continues to be ignored and discounted (Global Assessment Report, 2015).

Two innovative technologies that are helpful in national and international disaster management are DesInventar, a Disaster Information Management System, and EM-DAT, an international database that provides core data on natural and technological disasters from 1900 to the present, compiled from multiple sources, including UN agencies, non-governmental organizations, insurance companies, research institutes, and press agencies (UNISDR, 2017). DesInventar is a tool for generating National Disaster Inventories and constructing databases that capture information on damage, loss, and other effects of disasters. It enables the systematic analysis of disaster trends and their impacts, and the subsequent development of better prevention, mitigation and preparedness measures to reduce the impact of future disasters on the communities (UNISDR, 2017).

MENTAL HEALTH AND WELL-BEING

The threat and perception of climate change, including exposure to heat waves, vector-borne diseases, degraded air and water quality, extreme weather events, and fears about food security, create a cumulative negative impact on well-being and, for many, anxiety, distress, depression, and other mental health disorders.

In Chapter 8 of *The Impacts of Climate Change on Health in the United States* (Dodgen et al., 2016), the following key findings on mental health and well-being were highlighted:

- Exposure to climate-related disasters causes stress and serious mental health issues, including post-traumatic stress disorder (PTSD), depression, and anxiety. While the majority recover eventually, a significant proportion develop chronic psychological dysfunction.
- Certain groups of people—children, pregnant and post-partum women, the elderly, the economically disadvantaged, the homeless, people with preexisting mental illness, those living in areas most susceptible to climate change impacts, and first responders--are more vulnerable to the impacts of climate change and are at higher risk for developing mental health disorders.
- Many experience mental health consequences from the threat of climate change, perceptions about the impending impacts, and changes in the environment.

- Extreme heat increases risks of injury and death for those with mental illness, including the elderly and those on prescription medications that impair the body's ability to regulate temperature (Dodgen et al., 2016).

It is not surprising that the mental health consequences are greater for vulnerable populations, because they are disproportionately injured, displaced or killed by extreme weather events compared to other groups. Low-income housing is an area of particular concern in extreme weather events. Eight years after Hurricane Katrina devastated New Orleans, less than half of the public housing units had been rebuilt. Five years after Hurricane Ike destroyed 70 percent of the structures in Galveston, Texas, only 8 percent of the low-income public housing units had been rebuilt! Poor people often are permanently dislocated by weather disasters; their housing is not rebuilt and they have to search for affordable housing elsewhere, sometimes hundreds of miles away. Such glaring disparities must be addressed to achieve environmental justice. Counseling and support groups are part of the array of services needed to help people dislocated by a disaster to recover. In the U.S., the Substance Abuse and Mental Health Services Administration (SAMHSA) offers a free Disaster Distress Helpline – 1-800-985-5990. The Disaster Distress Helpline provides confidential crisis counseling and support 24/7 to people experiencing stress, anxiety, and other depression-like symptoms (SAMHSA, 2017).

Cultural competency is important in dealing with people dislocated by a natural disaster; people's responses to the stress of disasters vary considerably according to their beliefs, cultural traditions, and economic and social status. SAMHSA offers a webinar about how to utilize cultural awareness in disaster behavioral health planning (SAMHSA, 2017).

CONCLUSION

Rising greenhouse gas concentrations due to climate change have already caused rising temperatures, rising sea levels, changes in precipitation, and more frequent, intense extreme weather events (natural hazards). The impacts of climate change pose a significant threat to the health of humans and wildlife. Increases in contamination of food and water, increases in vector-borne diseases, degraded outdoor and indoor air quality, intense, prolonged heat waves, and extreme weather events cause illness, injury and death, destroy homes and property, and undermine the strides in development made by the poor.

To adapt to climate change one must first understand what it is. Health education about the impact of climate change on our health should begin in grade school and continue through adulthood at the workplace and in the community. People can be helped to remain safe by forming a community network and signing up for emergency response CodeRed. Warning systems must be multifaceted to reach even the most vulnerable. Color flags can be used on schools and other community organizations to notify residents of poor or dangerous air quality. Technologies can be used to instantly send alerts about flooding, heat waves, and other weather-related events by telephones, radio, television, and computer. At the same time, older non-digital technology such as flyers, bulletin boards, and newspapers should remain in use, as not everyone has access to digital technology, especially in low-income neighborhoods.

REFERENCES

Ahmed, S.M., Lopman, B.A., & Levy, K. (2013). A systematic review and meta-analysis of the global seasonality of norovirus. *PLoS ONE, 8*, e75922. doi:10.1371/journal.pone.0075922

American Lung Association. (2016). State of the Air 2016. Chicago, IL: ALA. Retrieved from http://www.lung.org/assets/documents/healthy-air/state-of-the-air/sota-2016-full.pdf

American Lung Association. (2017a). 10 Tips to Protect Yourself from Unhealthy Air. Download Phone APP on Outdoor Air Quality. Chicago, IL: ALA. Retrieved from http://www.lung.org/our-initiatives/healthy-air/outdoor/air-pollution/10-tips-to-protect-yourself.html

American Lung Association. (2017b). How to Know If Your Air Is Unhealthy. Chicago, IL: ALA. Retrieved from http://www.lung.org/our-initiatives/healthy-air/indoor/at-home/how-to-know-if-your-air-is-unhealthy.html

American Red Cross. (2017). How to treat heat-related illnesses. Retrieved from http://www.redcross.org/get-help/prepare-for-emergencies/types-of-emergencies/heat-wave-safety#Treat-Heat-Related-Illness

American Society of Heating, Refrigeration, and Air-conditioning engineers (ASHRAE). (2017). Standards and guidelines. Retrieved from https://www.ashrae.org/standards-research--technology/standards--guidelines

Astrom, D. O., Forsberg, B., & Rocklov, J. (2011). Heat wave impact on morbidity and mortality in the elderly population: A review of recent studies. *Maturitas*, *69*(2), 99–105. doi:10.1016/j.maturitas.2011.03.008 PMID:21477954

Basch, C. E. (2011). Executive summary: Healthier students are better learners. *The Journal of School Health*, *81*(10), 591–592. doi:10.1111/j.1746-1561.2011.00631.x PMID:21923869

Beard, C. B., Eisen, R. J., Barker, C. M., Garofalo, J. F., Hahn, M., & Shraamm, P. (2016). Vector-borne Diseases. In The Impacts of Climate Change on Human Health in the United States: A Scientific Assessment. Washington, D.C.: U.S. Global Change Research Program. Retrieved from https://health2016.globalchange.gov/

Berman, J. (2009). WHO: Waterborne disease is world's leading killer. Voice of America. Retrieved from http://www.voanews.com/a/a-13-2005-03-17-voa34-67381152/274768.html

British Broadcasting Channel (BBC). (2016). India records its hottest day ever. Retrieved from http://www.bbc.com/news/20039682

Buckley, W., & Piao, V. (2016, April 11). Rural Water, Not City Smog, May Be China's Pollution Nightmare. *The New York Times (Asia Pacific)*. Retrieved from https://www.nytimes.com/2016/04/12/world/asia/china-underground-water-pollution.html

Carter, L. M. et al.. (2014). Southeast and the Caribbean. In J. M. Melillo, T. Richmond, & G. W. Yohe (Eds.), *Climate Change Impacts in the United States: The Third National Climate Assessment* (pp. 396–417). Washington, D.C.: U.S. Global Change Research Program.

Cawley, C. (2016). Foobot indoor air quality review. Retrieved from http://www.makeuseof.com/tag/foobot-indoor-air-quality-monitor-review/

Centers for Disease Control and Prevention (CDC). (2014). Campylobacter. Atlanta, GA: US Department of Health and Social Services. Retrieved from https://www.cdc.gov/foodsafety/diseases/campylobacter/index/html

Centers for Disease Control and Prevention (CDC). (2015). Food Safety: Foodborne Germs and Illnesses. Atlanta, GA: Centers for Disease Control and Prevention. Retrieved from http://www.cdc.gov/foodsafety/foodborne-germs.html

Centers for Disease Control and Prevention (CDC). (2016). Division of Vector-Borne Diseases Fact Sheet. Atlanta, GA: U.S. Department of Health and Human Services. Retrieved from https://www.cdc.gov/ncezid/dvbd/pdf/dvbd_factsheet.pdf

Centers for Disease Control and Prevention (CDC). (2017a). West Nile Virus. Atlanta, GA: U.S. Department of Health and Human Services. Retrieved from https://www.cdc.gov/westnile/statsmaps/preliminarymapsdata/histatedate.html

Centers for Disease Control and Prevention (CDC). (2017b). Zika Virus. Atlanta, GA: U.S. Department of Health and Human Services. Retrieved from https://www.bing.com/news/search?q=cdc+zika+stats&qs=n&form=NWRFSH&pq=cdc+zika+stats&sc=8-14&sp=-1&sk

Chan, M. (2014). *Foreword. A global brief on vector-borne-diseases*. Geneva, Switzerland: WHO.

Confalonieri, U., Menne, B., Akhtar, R., Ebi, K. L., Hauengue, M., Kovats, R. S., & Woodward, A. et al. (2007). Human Health. In M. L. Parry, O. F. Canziani, J. P. Palutikof, P. J. van der Linden, & C. E. Hansson (Eds.), *Climate Change 2007: Impacts, Adaptation and Vulnerability. Contribution of Working Group II to the Fourth Assessment Report of the Intergovernmental Panel on Climate Change* (pp. 391–431). Cambridge, U.K.: Cambridge University Press.

Cook, B. I., Anchukaitis, K. J., Touchan, R., Meko, D. M., & Cook, E. R. (2016). Spatiotemporal drought variability in the Mediterranean over the last 900 years. *Journal of Geophysical Research*, *121*(5), 2060–2074.

CRED. (2015). Centre for research on the epidemiology of disasters: the international disaster database—EM-DAT. Retrieved from http://www.emdat.be/

Crescente, F. (2016, July 21). Obama signs Global Food Security Act to end hunger. *USA Today*. Retrieved from http://www.usatoday.com/story/news/politics/2016/07/20/obama-signs-global-food-security-act-to-end-hunger-invest-in-agriculture/87358640/

Dilley, M., Chen, R. S., Deichmann, U., Lerner-Lam, A. L., Arnold, M., & Yetman, G. et al. (2005). *Natural disaster hotspots: A global risk analysis*. Washington, D.C.: The World Bank Hazard Management Unit. doi:10.1596/0-8213-5930-4

Distel, O. (2016, June 27). Israel's water revolution and its overflow for the world. *Times of Israel Blog*. Retrieved from http://blogs.timesofisrael.com/israels-water-revolution-and-its-overflow-for-the-world/

Dodgen, D., Donato, D., Kelly, N., LaGreca, A., Morganstein, J., . . . Ursano, R. (2016). Mental Health and Well-Being. In The Impact of climate change on human health in the United States: a scientific assessment. Washington, D.C.: U.S. Global Change Research Program. Retrieved from https://health2016.globalchange.gov/

ENERGY STAR. (2017). ENERGY STAR certification, a joint program of the U.S. EPA and DOE. Washington, D.C.: U.S. Environmental Protection Agency. Retrieved from https://www.energystar.gov/

Food Assets for Assistance (FFA). (2017). World Food Program. Rome, Italy: World Food Program. Retrieved from http://www.wfp.org/food-assets

FoodSafety.gov. (2017). Check your steps. Washington, D.C.: Department of Health and Human Services. Retrieved from https://www.foodsafety.gov/keep/basics/index.html

Global Goals for Sustainable Development. (2017). Retrieved from http://www.globalgoals.org/

Grossman, M. M. (2016, June 28). Israel leading a "water revolution" in arid California. Jerusalem Post. Retrieved from http://www.jpost.com/Business-and- Innovation/Environment/Israel-leading-a-water-revolution-in-arid-California-457930

Hagens, W. I., & van Bruggen, M. (2015). Nationaal hitteplan. Retrieved from http://www.rivm.nl/Documenten_en_publicaties/Wetenschappelijk/Rapporten/2015/mei/

Kershner, I. (2015, June 1). Israel's water revolution. *Mosaic Magazine*. Retrieved from https://mosaicmagazine.com/picks/2015/06/israels-water-revolution/

Lal, A., Hales, S., French, N., & Baker, M.G. (2012). Seasonality in human zoonotic enteric diseases: A systematic review. *PLoS ONE, 7*, e31883. doi:10.1371/journal.pone.0031883

Lindsey, R. (2014). California facing worst drought on record. January 29, 2014. Climate.Gov: NOAA. Retrieved from https://www.climate.gov/news-features/event-tracker/california-facing-worst-drought-record

Loladze, I. (2014). Hidden shift of the ionome of plants exposed to elevated CO2 depletes minerals at the base of human nutrition. *eLife*, *3*, e02245. doi:10.7554/eLife.02245 PMID:24867639

Lowe, D., Ebi, K. L., & Forsberg, B. (2011). Heatwave early warning systems and adaptation advice to reduce human health consequences of heatwaves. *International Journal of Environmental Research and Public Health*, *8*(12), 4623–4648. doi:10.3390/ijerph8124623 PMID:22408593

Mengjie. (2015). Algae bloom expands in China's Yellow Sea. English.news.cn.: Qingdao, Shandong, China, June 4, 2015. Retrieved from http://news.xinhuanet.com/english/2015-06/04/c_134297379.htm

Mountain, T. C. (2016). Drought in Somalia: over 100 children die every day from starvation. Countercurrents.org. September 27, 2016. Available at: http://www.countercurrents.org/2016/09/27/drought-in-somalia-over-100-children-die-everyday-from-starvation

NASA. (2017). NASA, NOAA data show 2016 hottest year on record. Retrieved from https://www.nasa.gov/press-release/nasa-noaa-data-show-2016-warmest-year-on-record-globally

National Institute for Occupational Safety and Health (NIOSH). (2016). Health Hazards Evaluations (HHEs) Request. Atlanta, GA: Centers for Disease Control and Prevention. Retrieved from https://www.cdc.gov/niosh/hhe/request.html

National Oceanographic and Atmospheric Administration (NOAA). (2016). National Weather Service Heat Index. Silver Spring, MD: U.S. Department of Commerce. Retrieved from http://www.nws.noaa.gov/om/heat/heat_index.shtml

Occupational Safety and Health Administration (OSHA). (2016). *Safety and Health Topics: Indoor Air Quality*. Washington, D.C.: U.S. Department of Labor.

Omar, F. (2015). Unusual heat wave hits Kuwait, Middle East. *Kuwait Times*, September 8, 2015. Available at: ews.kuwaittimes.net/website/unusual-heat-wave-hits-kuwait-middle-east/

Organisation for Economic Co-operation and Development. (OECD (2014). The cost of air pollution: Health impacts of road transport. Retrieved from http://www.oecd.org/environment/ cost-of-air-pollution.htm

Organisation for Economic Co-operation and Development (OECD). (2016). The cost of air pollution. Paris: IECD Publishing Retrieved from http://www.oecd.org/env/the-cost-of-air-pollution-9789264210448-en.htm

Semenza, J. C., & Menne, B. (2009). Climate change and infectious diseases in Europe. Lancet ID, 9:365-375. Retrieved from http://ecdc.europa.eu/en/healthtopics/climate_change/health_effects/Pages/vector_borne_diseases.aspx#sthash.TNCMYLen.dpuf

Soller, J. A., Bartrand, T., Ashbolt, N. J., Ravenscraft, J., & Wade, T. J. (2010). Estimating the primary etiologic agents in recreational freshwaters impacted by human sources of faecal contamination. *Water Research*, *44*(16), 4736–4747. doi:10.1016/j.watres.2010.07.064 PMID:20728915

Spector, D. (2013, March 31). If you think China's air is bad, you should see its water. *Business Insider*. Retrieved from http://www.businessinsider.com/china-water-pollution-2013-3

Substance Abuse and Mental Health Services Administration (SAMHSA). (2017). Disaster preparedness, response, and recovery. Rockville, MD: SAMHSA. Retrieved from https://www.samhsa.gov/disaster-preparedness

Trtanj, J., Jantarasami, L., Brunkard, J., Collier, T., Jacobs, J., . . . Thurston, J. (2016). Climate impacts on water-related illness. In The Impacts of Climate Change on Human Health in the United States: A Scientific Assessment. Washington, D.C.: U.S. Global Change Research Program. Retrieved from https://health2016.globalchange.gov/

UNICEF. (2015). 500 children die every day from lack of safe water, sanitation in sub-Saharan Africa. New York: UNICEF. Retrieved from https://www.unicef.org/media/media_86521.html

UNICEF. (2016). Water and sanitation. New York: UNICEF. Retrieved from https://www.unicef.org/supply/index_cpe_water.html

Union of Concerned Scientists. (2017). Vehicles, air pollution, and human health. Retrieved from http://www.ucsusa.org/clean-vehicles/vehicles-air-pollution-and-human-health#.WKoYaYWcGhc

UNISDR. (2017). Disaster statistics. New York: United Nations. Retrieved from http://www.desinventar.net/index_www.html

United Nations. (2015). Paris Agreement. Retrieved from http://unfccc.int/files/essential_background/convention/application/pdf/english_paris_agreement.pdf

United Nations. (2017). Sustainable Development Goals: 17 goals to transform our world. Retrieved from http://www.un.org/sustainabledevelopment/hunger/

United Nations Environment Programme (UNEP). (2014). UNEP Yearbook 2014 emerging issues update. Air Pollution: World's Worst Environmental Health Risk. Retrieved from http://www.unep.org/yearbook/2014/PDF/chapt7.pdf

U.S. Environmental Protection Agency (EPA). (2014). *Energy savings plus health. Indoor air quality guidelines for school building upgrades*. Washington, D.C.: US EPA.

U.S. Environmental Protection Agency (EPA). (2016). Airnow. Air Quality Flag Program: Know your air quality to protect your health. Washington, D.C.: U.S. Environmental Protection Agency. Retrieved from https://airnow.gov/index.cfm?action=flag_program.index

U.S. Environmental Protection Agency (EPA). (2017a). Heat island compendium. Washington, D.C.: U.S. Environmental Protection Agency. Retrieved from https://www.epa.gov/heat-islands/heat-island-compendium

U.S. Environmental Protection Agency (EPA). (2017b). Particulate matter (PM) pollution. Washington, D.C.: U.S. Environmental Protection Agency. Retrieved from https://www.epa.gov/pm-pollution/particulate-matter-pm-basics#PM

USGCRP. (2016). The Impacts of Climate Change on Human Health in the United States: A Scientific Assessment (A. J. Crimmins, . . ., Eds.). Washington, D.C.: US Global Change Research Program; Retrieved from https://health2016.globalchange.gov/

WaterAid. (2016). Overflowing Cities: The State of the World's Toilets 2016. Retrieved from http://wateraidindia.in/publication/overflowing-cities-state-worlds-toilet-2016

WHO/UNICEF Joint Monitoring Programme for Water Supply and Sanitation. (2008). *Progress in Drinking Water and Sanitation: Special Focus on Sanitation. New York: UNICEF*. Geneva, Switzerland: WHO.

WHO/UNICEF Joint Monitoring Programme for Water Supply and Sanitation. (2015). New York: UNICEF. Geneva, Switzerland: WHO. Retrieved from http://www.wssinfo.org

World Bank. (2011). Wood-based biomass energy development for sub-Saharan Africa: Issues and approaches. Africa Renewable Energy Access Program. Retrieved from http://siteresources.worldbank.org/EXTAFRREGTOPENERGY/Resources/717305- 1266613906108/BiomassEnergyPaper_WEB_Zoomed75.pdf

World Health Organization. (2006). *WHO Air quality guidelines for particulate matter, ozone, nitrogen dioxide and sulfur dioxide*. Geneva, Switzerland: WHO.

World Health Organization. (2014a). Burden of disease from Household Air Pollution for 2012. Geneva, Switzerland: WHO. Retrieved from http://www.who.int/phe/health_topics/outdoorair/databases/HAP_BoD_results_March2014.pdf?ua=1

World Health Organization. (2014b). A Global Brief on Vector-borne diseases. Geneva, Switzerland: WHO. Retrieved from http://apps.who.int/iris/bitstream/10665/111008/1/WHO_DCO_WHD_2014.1_eng.pdf

World Health Organization. (2015). *WHO Initiative to Estimate the Global Burden of Foodborne Diseases*. Geneva, Switzerland: WHO.

World Health Organization. (2016a). Ambient (outdoor) air quality and health. Geneva, Switzerland: WHO. Retrieved from http://www.who.int/mediacentre/factsheets/fs313/en/

World Health Organization. (2016b). Fact sheet: Malaria. Geneva, Switzerland: World Health Organization. Retrieved from http://www.who.int/malaria/media/world-malaria-report-2016/en/

World Health Organization. (2017). WHO Global Urban Ambient Air Pollution Database (update 2016). Geneva, Switzerland: WHO. Retrieved from http://www.who.int/phe/health_topics/outdoorair/databases/cities/en/

Xu, Z., Sheffield, P., Su, H., Wang, X., Bi, Y., & Tong, S. (2014). The impact of heat waves on childrens health: A systematic review. *International Journal of Biometeorology*, *58*(2), 239–247. doi:10.1007/s00484-013-0655-x PMID:23525899

Ziska, L., Crimmins, A., Auclair, A., DeGrasse, S., Garofalo, J. F., Khan, A. S., & Walls, I. (2016). Food Safety, Nutrition, and Distribution. In *The Impacts of Climate Change on Human Health in the United States: A Scientific Assessment* (Ch. 7). Washington, D.C.: U.S. Global Change Research Program. Retrieved from https://health2016.globalchange.gov/

KEY TERMS AND DEFINITIONS

Carbon Sequestration: The process by which trees and plants absorb carbon dioxide, release the oxygen, and store the carbon.

Climate: The average weather conditions that persist for decades or longer.

Climate Change: Long-term changes in weather, including average temperature, amount of precipitation, and frequency of severe weather events.

Food Safety: Conditions and measures necessary for food production, processing, storage, and distribution in order to ensure a safe, sound, wholesome product that is fit for human consumption.

Foodborne Illness or Disease: Foodborne illness (sometimes called "food poisoning") caused by ingestion of food contaminated by pathogens, toxins, or chemicals.

Ground-Level Ozone: "Smog," an irritating gas created by chemical reactions between oxides of nitrogen (NOx) and volatile organic compounds (VOC) in the presence of sunlight, can cause health problems, particularly for children, the elderly, and people with asthma.

Heat Advisory: Issued within 12 hours of the onset of the following conditions: heat index of at least 105°F but less than 115°F for less than 3 hours per day, or nighttime lows above 80°F for 2 consecutive days (NOAA).

Heat Exhaustion: A mild form of heat stroke, characterized by faintness, dizziness, and heavy sweating.

Heat Index: The Heat Index (HI) or the "Apparent Temperature" is an accurate measure of how hot it really feels when the Relative Humidity (RH) is added to the actual air temperature.

Heat Island Effect: Phenomenon that results in urban areas being hotter than suburban and rural areas due to build-up of heat in asphalt and buildings and lack of shade or ground cover.

Heat Stroke: A condition resulting from excessive exposure to intense heat, characterized by high fever, collapse, and sometimes convulsions or coma.

Heat Wave: A period of abnormally and uncomfortably hot and unusually humid weather, typically lasting two or more days.

Natural Hazard: A naturally occurring extreme event such as a hurricane, earthquake, tornado, or volcanic eruption.

Vector: An insect, tick or other organism that transmits a disease from a plant or animal to an animal by biting it.

Chapter 3
Identifying and Mapping Vulnerable Populations

ABSTRACT

Climate change does not affect all populations uniformly; there are disproportionate impacts on certain populations who are more vulnerable to displacement, injury or death due to risk factors such as poverty, race, age, and medical conditions. The social determinants of health interact with climate change to determine a population's risk of adverse health impacts from changing precipitation patterns, rising temperatures, rising sea levels, and extreme weather events. Emergency preparedness planning must take into account the special concerns of vulnerable populations in order to ensure that they have food, water, and a temperature-safe shelter during climate change events such as heat waves, floods, and hurricanes. In addition, they may require assistance in evacuating their residences. This chapter addresses the use of technologies to identify and map these vulnerable populations.

INTRODUCTION: DEFINING VULNERABLE POPULATIONS

The first step in identifying populations of concern is to define "vulnerability" and how it applies to public health impacts in the context of global warming.

Merriam Webster's Dictionary defines "vulnerability" as "the quality or state of having little resistance to some outside agent (vulnerability to infection*)*" (Merriam Webster, 2017). In the context of climate change threats to health, vulnerability is the tendency or predisposition to be adversely affected

DOI: 10.4018/978-1-5225-3414-3.ch003

by these threats. Vulnerability to climate-related health effects comprises three elements: exposure, susceptibility to harm, and adaptive capacity. In the Introduction to *The Impacts of Climate Change on Human Health in the United States: A Scientific Assessment,* Balbus et al. (2016) define the three elements as follows:

- Exposure is contact between a person and one or more biological, psychosocial, chemical, or physical stressors, including stressors affected by climate change. Contact may occur in a single instance or repeatedly over time, and may occur in one location or over a wider geographic area.
- Susceptibility to harm (or sensitivity) is the degree to which people or communities are affected, either adversely or beneficially, by climate variability or change.
- Adaptive capacity is the ability of communities, institutions, or people to adjust to potential hazards, to take advantage of opportunities, or to respond to consequences. A related term, resilience, is the ability to prepare and plan for, absorb, recover from, and more successfully adapt to adverse events (Balbus et al., 2016, adapted from IPCC, 2014 and NRC, 2012).

Vulnerability occurs on an individual level, affected by the person's health status (e.g., presence of underlying medical condition such as asthma), age, gender, educational level, income, family support, and access to healthcare. It also occurs at the societal level, where adaptive capacity is influenced strongly by the natural and built environments (for example, the infrastructure), governance and management (health-protective surveillance programs, regulations and enforcement, or community health programs), and institutions. Exposure to extreme weather events is in part a function of geography, (e.g., coastal areas and islands are especially at risk of storms and flooding). Exposure to pathogens is a function of location (near vector habitats) but can be greatly reduced through integrated pest management. Susceptibility to harm, or sensitivity, may be increased by a pre-existing condition (e.g., asthma or other lung disease), age (i.e., very young or elderly), pregnancy, occupation. The greatest predisposing factor for vulnerability is poverty. Racial, religious, ethnic, or gender discrimination is another risk factor for vulnerability.

Maplethorpe's Climate Change Vulnerability Index 2016 ranks 186 nations' vulnerability to the adverse impacts of climate change according to their risk for exposure (50 percent), sensitivity (25 percent), and adaptive capacity (25 percent). The 2016 Index deemed Chad as the country most vulnerable to the impacts of global warming; the second most vulnerable country was Bangladesh. The 2016 Index rated Africa as the most vulnerable region because it is exceptionally exposed to unpredictable rainfall patterns and prolonged drought, suffers from extreme levels of poverty, hunger, and lack of clean water, and is racked by conflicts and an inability to implement climate adaptation policies. The second most vulnerable regions are Central America and Asia (Ranada, 2016).

In Chapter 9 of *The Impacts of Climate Change on Human Health in the United States: A Scientific Assessment* (Gamble et al., 2016), the following key findings are noted:

- Vulnerability varies over time and is place-specific.
- Health impacts vary with age and life stage, with the most vulnerable being the very young and the very old.
- Social determinants of health interact with climate factors to affect health risks; for example, socioeconomic factors and health disparities have a strong influence on health outcomes.
- Mapping tools and vulnerability indices identify climate health risks.

The Impacts of Climate Change on Human Health in the United States: A Scientific Assessment devoted an entire chapter to the discussion of vulnerable populations, or "populations of concern"; the following eight groups were identified (Gamble et al., 2016):

- Communities with environmental justice concerns (communities of color, low income groups, immigrants, and limited English proficiency groups);
- Indigenous peoples;
- Pregnant women;
- Children;
- Older adults;
- Occupational groups;
- People with pre-existing health conditions;
- People with disabilities.

The following discussion examines each of these vulnerable populations.

Communities With Environmental Justice Concerns

The U.S. Environmental Protection Agency (2017) defines environmental justice (EJ) as "the fair treatment and meaningful involvement of all people regardless of race, color, national origin, or income, with respect to the development, implementation, and enforcement of environmental laws, regulations, and policies." Fair treatment means that there should be no health disparities, that no group should bear a disproportionate share of the negative environmental consequences resulting from industrial, governmental and commercial operations or policies. Meaningful involvement means that all people will have an opportunity to participate in decisions about activities that may affect their environment and/or health, that decision makers will seek out and facilitate community involvement in the decision-making process. Environmental justice will be achieved when everyone enjoys equal protection from environmental and health hazards, and equal access to the decision-making process of creating a healthy environment in which to live, learn, and work. (See Chapter 4 for a more in-depth discussion of environmental justice and climate change.)

Populations with EJ concerns include communities of color, low income groups, certain immigrants, and limited English proficiency (LEP) groups. These populations are at an increased risk of exposure because they typically live in risk-prone areas (e.g., flood-prone areas, isolated rural areas, urban heat islands) with poorly maintained infrastructure and an increased burden of air pollution (CDC, 2011; Frey, 2011; Luber & McGeehin, 2008; Miranda et al., 2011). They also have a greater incidence of chronic medical conditions than do other non-vulnerable groups, and tend to live in lower SES neighborhoods with limited transportation, limited access to health education, and limited medical care; in addition, lack of education and language deficiencies impair their ability to cope with climate-change related health impacts (DHHS, 2014; Knowlton et al., 2009; McMichael, 2013; Satia, 2009; Semenza et al., 1999; Younger et al., 2008). For undocumented immigrants and LEP populations, high poverty rates, language and cultural barriers, and lack of citizenship limit access to and use of health care and social services, as these groups are afraid of being detained and deported (Eneriz-Wiemer et al., 2014; Fuentes-Afflick & Hessol, 2009; Maldonado et al., 2013; Ortega et al., 2007; Vargas Bustamente et al., 2010). People of color--Hispanics, Blacks/African

Americans, American Indians and Alaska Natives, Asian Americans, and Native Hawaiians and Pacific Islanders--already comprise 37 percent of the total U.S. population and by 2042 are projected to be in the majority (Humes et al., 2011; U.S. Census Bureau, 2008, 2014a).

Health impacts of climate change affecting communities of color, immigrants, LEP groups, and low-income groups include heat waves and other extreme weather events, poor air quality, food safety, infectious diseases, and psychological stressors. Heat waves disproportionately affect these groups because they live in areas subject to heat island effects and have reduced adaptive capacity due to inadequately insulated housing, lack of air conditioning, inadequate access to public cooling centers, and inadequate access to both routine and emergency health care (CDC, 2011; CDC, 2013; Luber & McGeehin, 2008). (See Chapter 7 for discussion of the 1995 Chicago heat wave.) Losses of life and homes from natural disasters such as Hurricane Katrina in New Orleans in 2005 and the great Sichuan earthquake of 2008 disproportionately affect low-income communities of color because of their poor-quality housing, lack of access to emergency communications, lack of access to transportation, inadequate access to health care services and medications, limited post-disaster employment, and limited or no health and property insurance (Aiming, 2009; Arrieta et al., 2009; Donner & Rodriguez, 2008; Eisenman et al., 2007; Joseph et al., 2014; Lane et al., 2013; Otani et al., 2012; Zoroaster, 2010; Zhou et al., 2012).

Indigenous Peoples

Indigenous people are at high risk for poor mental health related to historical or personal trauma, alcohol abuse, suicide, infant/child mortality, environmental exposures from pollutants or toxic substances, and diabetes caused by inadequate or improper diets (Evans-Campbell, 2008; Hoover et al., 2012; Milburn, 2004; Sarche & Spicer, 2008; Wong et al., 2014.) These underlying health disparities predispose indigenous people, especially those who are dependent on the environment for sustenance or who live in isolated or impoverished communities, to greater exposure and lower resilience to climate-change-related health effects. Climate change affects indigenous communities' food safety and security, such as a reduction in the abundance and nutritional quality of berries in Alaska Native communities, declines in traditional rice harvests and moose populations (two food sources) in the Upper Great Lakes Ojibwe communities, and a drought among Navajo people

(Cheruvelil & Barton, 2013; Dybas, 2009; Kellogg et al., 2010; Lenarz et al., 2009; Nakashima et al., 2012; Redsteer et al., 2011).

Indigenous communities who rely on fishing are impacted by changes in the aquatic habitats. Oceans are becoming more acidic as they absorb more CO_2, and the increased acidity adversely impacts shellfish survival. Reduced oxygen levels in freshwater and seawater promote the growth of disease-causing bacteria, viruses, and parasites, and degrade water quality (Melillo et al., 2014). Rising temperatures exacerbate shellfish disease and are associated with increased accumulation of methylmercury in fish and increased human exposure to mercury, a dangerous neurotoxin. Exposure to mercury is particularly dangerous at the prenatal stage (Booth & Zeller, 2005; Cozetto et al., 2013; Dijkstra et al, 2013). Access to clean drinking water and sanitation also are issues for a significant number of indigenous peoples on remote reservations that lack indoor plumbing and rely on unregulated water supplies that are vulnerable to drought, changes in water quality, and contamination of water in local systems (Redsteer et al, 2011; Tribal Water Working Group, 2012; US EPA, 2008).

Climate change threatens the cultural identity of indigenous peoples by reducing the availability of plant and animal species used in sacred ceremonial and cultural practices (Doyle et al., 2013). The loss of medicinal plants may leave medicine men and women without the resources they need to practice traditional healing (Lynn et al., 2013; Voggesser et al., 2013). Climate-change impacts compound the profound losses already experienced historically under colonialism, when the loss of tribal lands and the suppression of tribal languages and culture eroded cultural identity (Cunsolo Willox, 2012; Cunsolo Willox et al, 2013; Cunsolo Willox et al., 2015; Hodge et al., 2009; Hodge & Limb, 2010). Many young people are leaving the reservations to seek education and work opportunities elsewhere, which may reduce intergenerational interactions and undermine the sharing of traditional knowledge, tribal lore, and oral history (Alessa et al., 2008).

Pregnant Women

Pregnant women and the developing fetus are the most vulnerable humans (Rylander et al., 2013). As a result, the negative impacts of climate change are especially hazardous for maternal health, posing severe health risks to both the mother and the fetus. Low birth weight, preterm birth, and congenital cataracts have been linked to exposure to heat waves, airborne particulate matter, and floods (Gamble et al., 2016; van Zutphen et al., 2012). Direct heat exposure, such as in heat waves, threatens the health of mother and the child. Newborns are especially sensitive to extreme temperatures because they have a limited capacity to regulate body temperature (Polin & Abman, 2011).

In addition, heat waves contribute to foodborne diseases and contamination (see Chapter 2). Extreme weather events such as droughts lead to crop failure, livestock mortality, and increased cereal prices, which cause food shortages and malnutrition. A malnourished mother has fewer nutrients available for the growing fetus. Maternal weight gain affects fetal growth. Low weight gains during pregnancy are associated with an increased risk of preterm birth and low birth weight (MedNet 2010). Another risk for pregnant women and the unborn fetus is the lack of safe drinking water. Safe drinking water is unavailable to more than 1 billion people worldwide (Akachi et al., 2009). Natural disasters will decrease access to safe drinking water and sanitation, increasing the risk of malnutrition, diarrhea, and cholera. Rising temperatures and sea levels have contributed to degrading the quality of freshwater and seawater. Climate change impacts are projected to increase the population of food-insecure countries to 6.8 billion, or approximately 80 percent of the world's population (Fischer et al., 2005).

In addition, exposure of pregnant women to inhaled particulate matter is associated with negative birth outcomes (Choi et al., 2008; Makri & Stilianakis, 2008; Jayachandran, 2009; Ritz et al., 2007). Incidences of diarrheal diseases and dehydration may increase, with adverse impacts on pregnancy outcomes and newborn health (Rylander et al., 2013). Floods increase the risk of maternal exposure to environmental toxins and mold, reduce the mother's access to safe food and water, create psychological stress, and disrupt prenatal health care. Floods also increase the maternal risk of anemia (iron deficiency), eclampsia (a condition that can cause seizures in pregnant women), and spontaneous abortion (Callaghan et al., 2007; Tees et al., 2010; Hamilton et al., 2009; Harville et al., 2009).

Approximately 536,000 women die annually from complications during pregnancy, childbirth, or the 6 weeks following delivery (Homer et al., 2009). Half of all maternal deaths are from blood loss or hypertension. Other indirect causes, such as malaria, HIV/AIDS, and heart diseases, account for 18 percent of maternal deaths. Most maternal deaths are avoidable, and 99 percent of all maternal deaths related to childbearing or giving birth occur in developing countries (Costello et al., 2009).

Children

Children are more vulnerable than adults to adverse health effects associated with environmental exposures due to a number of factors; first, they take in more oxygen, water and food per pound of their body weight than do adults, so, for example, when breathing in polluted air, their lungs are exposed to a greater concentration of pollutants (Shannon et al., 2007). Second, their exposure is increased by the nature of their interaction with the environment; young children often play on the floor indoors or on the ground outside and they tend to put their fingers and objects in their mouths, increasing their exposure to dust, pesticides, mold spores, and allergens (Roberts et al., 2009).

Climate change increases children's exposure to health threats. Climate-related events interact with the child's stage of development as well as with socioeconomic factors such as economic status, diet, and living situation. Children living in poverty are hard-hit by climate change. Poor and low-income households have difficulty accessing health care and meeting the basic childhood needs for healthy development. Children in poverty are less likely to have access to air conditioning and to be able to respond to or escape from extreme weather events (Balbus & Malina, 2009; Sheffield & Landrigan, 2011; Bernstein & Myers, 2011; Xu, Sheffield et al., 2012; Xu, Etzel et al., 2012).

Increasingly frequent and intense heat waves are already taking their toll on children and teenagers, especially student athletes and others who spend times outdoors or in indoor non-climate-controlled rooms. Approximately 9,000 high school athletes in the United States are treated for exertional heat illness (such as heat stroke and muscle cramps) each year (Gilchrist et al., 2010; Kerr et al., 2013). From 1997 to 2006, emergency department visits for all heat-related illness increased 133 percent and youth made up almost 50 percent of those cases (Nelson et al., 2011). From 2000 through 2013, heat stroke fatalities doubled among U.S. high school and college football players (Gottschalk & Andrish, 2011).

Extreme weather events have traumatic effects on children. In 2003, more than 10 percent of U.S. children from infancy to 18 years of age reported experiencing a disaster (fire, tornado, flood, hurricane, earthquake, etc.) during their lifetimes (Becker-Blease et al., 2010). Living through a disaster can impact children's capacity to regulate emotions, undermine their cognitive development and academic performance, and lead to mental health disorder such as post-traumatic stress disorder (PTSD), depression, anxiety, phobia, and panic disorders (Fairbank et al., 2014). As with adults, children's adaptive ability is affected by socioeconomic status and available support systems. If mental health disorders that develop in children are left untreated, they can extend into adulthood (Fairbank et al., 2014).

Degraded outdoor air quality due to ground-level ozone and particulate matter is associated with increases in asthma episodes and other adverse respiratory effects in children (Strickland et al., 2010; Ostro et al., 2009; Parker et al., 2009). About one in 11 children in the United States suffer from asthma; it is a leading cause of school absenteeism in the United States (CDC, 2017). Rising temperatures from climate change create longer flowering seasons, which increase the number of airborne allergens, leading to increased allergy and asthma symptoms. Children may also be exposed to indoor air pollutants such as tobacco smoke and mold.

Climate change has been increasing waterborne infections (see Chapter 2), and children are particularly susceptible because they swallow about twice as much water as adults when swimming. An increased association between heavy rainfall and increased acute gastrointestinal illness has already been observed in children in the United States (Drayna et al., 2010). From 1997 to 2006, children accounted for 40 percent of swimming-related eye and ear infections from the waterborne bacteria *Vibrio alginolyticus* (Dechet et al., 2008); in addition, in 2009-2010, 66 percent of people seeking treatment for illness associated with harmful algae bloom toxins were youths aged 1-19 (Hilborn et al., 2014).

Climate change has also increased vector-borne infections, and children are particularly at risk of acquiring or having complications from listeriosis (Janakiraman, 2008), dengue fever (Chitra & Panicker, 2011), and influenza (Rasmussen et al., 2012). Children spend more time outdoors than adults do, which increases their exposure to mosquito and tick bites that can cause La Crosse encephalitis or Lyme disease (Sheffield & Landrigan, 2011; Xu, Sheffield et al., 2012; Erwin et al., 2002). Lyme disease trended upward in children from 1992 to 2006 (Ebi et al., 2008; CDC, 2015a). Young children are especially vulnerable to malaria; in 2015, 69 percent of all malaria deaths worldwide were in children under five years of age (WHO, 2016).

Older Adults

Older adults (those 65 years or older) are vulnerable to the health impacts of climate change and extreme weather (Balbus & Malina, 2009; Filiberto et al.,

2010; Wang et al., 2010). The nation's older adult population is projected to increase from approximately 48 million in 2015 to 88 million in 2050 (U.S. Census Bureau, 2010). Of those 88 million older adults, nearly 19 million will be 85 years of age and older in 2050 (U.S. Census Bureau, 2014b). Older adults are not a homogeneous group, but rather, comprise a number of subgroups based on race, educational attainment, socioeconomic status, social support networks, physical and mental health, and disability status. Older adults are more vulnerable to multiple adverse health consequences from exposure to extreme heat (O'Neill & Ebi, 2009). In New York City, extreme high temperatures were associated with increased hospital admissions of the elderly for cardiovascular and respiratory disorders. Hospital admissions for older adults increased 4.7 percent per degree Celsius increase (Lin et al., 2009). Diabetes, which increases the risk of adverse consequences from exposure to intense heat, affects more than 25 percent of the older adult population (Kirkman et al., 2012).

Extreme weather events that require evacuation pose increased health and safety risks for older adults, especially those living in poverty or in nursing or assisted-living facilities. Moving fragile elderly patients to a shelter is expensive, time-consuming and requires transfer of medical records, medications, and medical equipment (Laditka et al., 2008; Little et al., 2004).

Air pollution poses a threat for older adults, especially those with breathing difficulties, asthma, or COPD. Exposure to air pollution increases the risk of a heart attack, especially for older adults who are diabetic or obese. Exposure to ground-level ozone can affect lung function and increase emergency department visits and hospital admissions, even for healthy older adults.

Some vector-borne diseases, especially mosquito-borne West Nile and St. Louis encephalitis viruses (Lindsey et al., 2010) pose a greater health risk among sensitive older adults with compromised immune systems. Older adults also are at greater risk for contracting gastrointestinal illnesses from contaminated drinking and recreational water and suffering severe health outcomes and death (Bush et al., 2014; Jagai et al., 2012; Naumova, 2003).

Occupational Groups

Climate change may increase the frequency and severity of known occupational hazards, as well as creating some new ones. Some groups of workers are at greater risk of adverse consequences from the health impacts of climate change because of the nature of their work. Climate change is expected to affect the health of outdoor workers by increasing mean temperatures, degrading outdoor air quality, creating more frequent and severe extreme weather events, and increasing the prevalence and distribution of vector-borne diseases, and industrial exposures (Schulte & Chun, 2009). Employees who work outdoors are heavily exposed to the impact of outdoor air pollution, heat waves, blizzards and others extreme weather events. Following is a list of some occupational groups whose employees are heavily exposed to outdoor air pollution and other outdoor effects of climate change:

- Farmers, ranchers, and other agricultural workers;
- Commercial fishermen;
- Construction workers;
- Paramedics, firefighters, police, and other first responders;
- Transportation workers;
- Members of the military (U.S. EPA, 2016).

Agricultural workers face the additional risks of being exposed to pesticides. For day laborers and migrant workers, the adverse health effects are compounded by the housing of the workers, which often lacks proper insulation, air conditioning, and hygiene. Firefighters are exposed to chemical contaminants released from burning buildings, and to the smoke from wildfires out in the field.

Other occupational groups at increased risk of exposure to the adverse effects of climate change include indoor workers who work in hot indoor environments such as steel mills, dry cleaners, manufacturing facilities, warehouses, and other places that lack air conditioning. The combination of exposure to extreme heat and exposure to VOCs and other indoor air contaminants poses a serious risk for these workers (U.S. EPA, 2016). Proper ventilation of indoor environments, ongoing monitoring of air quality, and enforcement of policies for pollution controls in such work environments are essential to workers' health and safety.

It is vital that employers understand the impacts of climate change so they can take steps to prepare and protect their workers, including educating them about the hazards, holding emergency drills, and enforcing safety codes. Employers, safety professionals, and workers need to remain informed about emerging issues and hazards associated with climate change to continue to develop plans that address worker safety and health.

People With Asthma and Other Chronic Diseases

Climate change is creating warmer temperatures and more frequent heat waves, which, in turn, lead to increases in heat stress, exacerbations of respiratory illnesses, and heat-related deaths. Individuals with asthma are at risk of worsened symptoms and asthma attacks from degraded air quality (Carter et al., 2014). Rising temperatures degrade outdoor air quality by increasing the concentration of ground-level ozone, which pollutes the air. When ozone remains in the upper atmosphere (the stratosphere), it shields us from ultraviolet radiation. However, at ground level, ozone creates smog, which can cause breathing problems by damaging lung tissue, reducing lung function, and inflaming airways (Carter et al., 2014). In addition, global warming increases the concentration of airborne plant allergens. Flowering starts earlier, resulting in a longer ragweed pollen season, and more time for exposure to airborne allergens (USGCRP, 2016). Those with allergic asthma may need to have medication adjusted.

Climate change increases the risk of storm surges and flooding, as well as king tides that create unsafe water conditions. The dampness from flooding will lead to the growth of mold, which is an asthma and allergy trigger. It is important to keep homes dry to avoid the buildup of mold. Once mold appears, it needs to be removed safely, using clean water (the authorities will notify residents when water is safe after a flood; contaminated water should not be used to clean walls or countertops) and disinfectants. People with asthma and other lung diseases should not use bleach in cleaning up after a flood, as the vapors from it can inflame the airways. The U.S. Environmental Protection Agency has a list of approved bleach-free disinfectants that are safe for households that include one or more persons with asthma or other lung diseases. If possible, keep the individuals with asthma away from the mold and let others do the cleaning, as exposure to mold may trigger an asthma attack. The CDC (2016) lists eight tips to stay healthy in cleaning up mold:

- Protect yourself—with gloves, mask, goggles; cover skin (long sleeves and pants).
- Toss—anything that is wet with floodwater that cannot be cleaned and dried within 24-48 hours should be discarded outside; take pictures for insurance.
- Air out your home; open doors and windows while you are cleaning.

- Circulate air—when it is safe to use electricity, use fans and dehumidifiers to remove moisture.
- Do not mix cleaning products.
- Scrub surfaces with clean water and detergent; dry right away.
- Do not cover mold—remove it.
- Dry your home and everything in it as quickly as possible—within 24-48 hours.

Climate change increases extreme heat waves and wildfires, which release particle pollution in smoke that can blow hundreds of miles away. This particle pollution is particularly dangerous for those with asthma, COPD, and heart disease; exposure to it can trigger asthma flare ups, coughing, and even heart attacks (Luber et al., 2014).

Individuals who take certain drugs for medical conditions may be at particular risk of succumbing to illness or death from heat waves. These medicines include drugs used to treat neurologic or psychiatric conditions, such as anti-psychotic drugs, anti-cholinergic agents, anxiolytics (anti-anxiety medicines), and some antidepressants (e.g., selective serotonin reuptake inhibitors or SSRIs), as well as diuretics and beta blockers used to treat cardiovascular disease (Martin-Latry et al., 2007; Stöllberger et al., 2009; Nordon et al., 2009; Hausfater et al., 2009).

In developing countries, the use of biomass as a cooking fuel has exposed women to unhealthy levels of indoor particulate matter, which has compromised their lung functioning. Biomass burning on cook stoves has been associated with the development of chronic bronchitis and obstructive airway disease. In rural Mexico, researchers found that women who used biomass as a cooking fuel had increased respiratory symptoms and reduced lung function compared with women who cooked with gas (Instituto Nacional de Enfermedades Respiratorias, 2006). Biomass burning in cookstoves emits black carbon, which absorbs solar energy and is a major contributor to global warming. Developing countries in Asia, Africa, and Latin America emit more than three-quarters of global black carbon emissions, from cookstoves and the burning of coal and wood for heating (Cho, 2016). The development and distribution of cleaner, more efficient cookstoves is one strategy to reduce the harm of the cookstoves (AFREA, 2011).

People With Disabilities

Climate change is increasing the frequency and severity of extreme weather events, which pose special challenges for those with a disability. A disability is any impairment of the body or mind that limits a person's ability to perform certain activities or participate in life activities such as school, work, and recreation. The functional limitations caused by disability include hearing, speech, vision, cognition, and mobility (CDC, 2015b). Disability varies by gender, race, ethnicity and geographic location (National Center for Health Statistics, 2014). Approximately 18.7 percent of the U.S. population has a disability. Although disability can occur at any age, it is far more prevalent among older adults; in 2010, in the United States, 16.6 percent of adults aged 18-64 had a disability, compared with a whopping 49.8 percent of those 65 years and older (Brault, 2012). Worldwide, an estimated one billion individuals are living with disabilities, with about 200 million of them experiencing substantial difficulties in functioning (WHO, 2011).

Individuals with disabilities have unemployment rates double those of non-disabled individuals, and are at an increased risk of poverty and lower educational achievement, two social risk factors that lead to poorer health outcomes during extreme weather events (Bureau of Labor Statistics, 2015). Without careful emergency planning that addresses the needs of those with disabilities, they can be expected to fare far worse than non-disabled individuals in extreme weather events. Warnings to boil water, evacuate, etc. need to be made accessible to hearing- and vision-impaired individuals. Functional impairments make evacuation before an impending extreme weather event challenging, as many individuals with disabilities need help in getting out of their homes and being transported to a shelter. In addition, their medications and equipment—wheelchair, oxygen, and so forth—must be transported with them. Hurricane Katrina, which hit New Orleans in 2005, and Superstorm Sandy, which struck New York in 2012, are two cases in point. Of 986 fatalities directly attributable to Hurricane Katrina, 103 were individuals in nursing homes (Brunkard et al., 2008). After Superstorm Sandy, the Brooklyn Center for Independence of the Disabled (Brooklyn Center for Independence of the Disabled et al. v. Bloomberg et al., Case 1.11-cv06690-JMF 2013) won a lawsuit against the City of New York; the court ruled that the city had not adequately prepared to accommodate the social and medical support needs of individuals with disabilities.

Strong social networks (social cohesion—see Chapter 7), especially through faith-based organizations, family networks, and work connections, were considered key enabling factors that helped people with disabilities to cope before, during, and after extreme weather events such as Hurricane Katrina and the Chicago Heat Wave of 1995 (Fox et al., 2010; Klinenberg, 2002, 2015).

EMERGENCY PREPAREDNESS: CREATING A DATABASE FOR VULNERABLE POPULATIONS

Identifying vulnerable, or at-risk, populations before an emergency will make evacuation much easier to accomplish. In fact, maintaining an up-to-date database of vulnerable populations aids in managing all four phases of a disaster, from preparedness, to response, recovery and mitigation. Two approaches are advocated—an individual approach (registries and COINs) and a population approach (social vulnerability index).

Registries may be useful in identifying vulnerable populations. Most people are familiar with bridal registries. In the context of emergency management, the Centers for Disease Control and Management (CDC) defines a registry as "a voluntary database of individuals who meet the eligibility requirements for receiving additional emergency response services based on needs" (CDC, 2015c, p. 5). Of course, there may be resistance to registering; emergency managers would do well to conduct their registry outreach through trusted community organizations. Information may be collected by volunteers or social service workers when clients apply for other public health services. Other methods of collecting information for a registry include signing up by phone, on line, or by having a form mailed to them with a postage-paid addressed envelope for easier return. Social media, TV, newspapers, and brochures can all be used to recruit vulnerable populations for a registry. In addition, tabling events may be used to recruit people for an at-risk registry; for example, a registry table could be set up at a local wellness or health fair or at ethnic celebrations or festivals in the community. Registries need to be updated annually, as people move, die, or their vulnerability status changes.

Another method of identifying vulnerable populations is to form a Community Outreach Information Network (COIN). A COIN is a grassroots network of trusted community leaders who can be helpful in defining, locating, and reaching at-risk individuals in an emergency. The Centers for Disease

Control and Management (2010) has published a public health workbook to aid emergency managers who wish to form a COIN. The first step of the COIN's job is to define the vulnerable groups in the community. What needs do these individuals have in terms of spoken languages, cultural practices, beliefs, and physical or mental limitations? The U.S. Census (http://www.census.gov/) and local health departments are recommended sources of useful information. The second step is to locate the at-risk individuals. To facilitate this process, identify any organizations that are working with this population and check out what they have already done. Partner with key community organizations and service providers; identify contacts within each of these groups and meet with them to discuss the importance of identifying at-risk groups during an emergency and what role each organization can play during an emergency. Nontraditional organizations such as local nonprofits and faith-based organizations may aid in reaching isolated at-risk individuals (CDC, 2015c). Conduct a survey or focus group with members of different at-risk groups to determine preferred method of communication, assistive technology used, use of media, primary languages spoken, and trusted people and sources of public health information. This process will aid the emergency management team in developing culturally competent messages and in delivering them in a way that they will be received and understood.

In a population approach, U.S. Census statistics and other databases are assessed to estimate the number of individuals in various at-risk categories. The CDC has developed a Social Vulnerability Index (SVI), a free on-line tool (https://svi.cdc.gov/) that uses national-level data. The SVI may be modified for each state or region. Social vulnerability refers to a community's resilience in the face of natural or human-caused disasters or disease outbreaks. Cultivating community resilience reduces social vulnerability and, thereby, human suffering and economic loss. This is achieved by enhancing a community's "capacity to anticipate, cope with, resist, and recover from the impact" of a disaster in nature or society (Chen et al., 2009). Social vulnerability to disaster is influenced by six main factors, namely, socioeconomic status (SES), age, gender, race and ethnicity, English language proficiency, and medical issues and disability (Flanagan et al., 2011; Keim, 2008). There is overlap among the categories. For example, someone with mental illness may have low SES, an elderly individual is more likely to have medical issues, and so on. SES consists of employment, housing, income and education; in a disaster such as storm flooding, a lower SES individual is less likely to be able to access food, be able to afford to miss work, or have resources to evacuate and stay outside the home for a period of time. Women, who, on average, make less money than men do, are more vulnerable than men. Children are more vulnerable than adults, as they are less likely to know how to cope with an emergency. In addition, their immune system is less developed than an adult's is, their skin is more sensitive, and they breathe in more air per pound than do adults and thus are more severely impacted by airborne toxins than are adults. Race and ethnicity contribute to SES; some groups historically have been marginalized and still suffer from disparities in lifestyle and health status. Individuals with limited English proficiency (LEP) have difficulty understanding, speaking, and reading English. In an emergency, instructions must be in more than one language if the residents speak a language other than English. In addition, evacuation procedures must be culturally sensitive, respecting the customs (e.g., clothing, modesty) of the different ethnic groups. Individuals with disabilities need to bring any assistive devices (e.g., wheelchair, walker, oxygen) and medications they use.

By identifying the location and needs of at-risk individuals in advance, emergency managers will be better able to plan an evacuation for those with special needs. Some at-risk groups lack transportation. Others with physical or cognitive disabilities may require assistance in evacuating their home.

MAPPING VULNERABLE POPULATIONS

The use of new technologies to visually represent geographic data allows for more sophisticated mapping of risk factors and social vulnerabilities to identify and protect specific locations and groups of people. Geographic Information Systems (GIS) are mapping tools that facilitate the visual representation and analysis of census data, data on the determinants of health (social, environmental, preexisting health conditions), measures of adaptive capacity (such as health care accessibility), and environmental data for the identification of at-risk populations (CSDH, 2008; Manangan et al., 2014; Friel et al., 2008). Mapping the relative vulnerabilities of different populations helps planners to identify in which regions and for whom climate health risks are greatest. As the efficacy and precision of the GIS mapping tools improve, scientists will be able to predict the probability of specific health impacts occurring in specific regions (Gamble et al., 2016).

CONCLUSION

In planning for climate change adaptation, it is vital to identify vulnerable populations in order to give them the support and protection they need to remain safe during extreme weather events. A population's vulnerability is a function of its exposure, susceptibility to harm, and adaptive capacity. These factors, in turn, are determined by geographic location, as well as by the social determinants of health, which underpin a community's ability to cope with challenging climate events such as floods or droughts. Vulnerable populations include communities with EJ concerns, indigenous peoples, pregnant women, children, older adults, certain occupational groups such as outdoor workers, people with asthma and other health conditions, and people with disabilities.

In developing a comprehensive emergency management plan, databases of vulnerable populations need to be compiled; an individual approach utilizes registries and COINs. A population approach uses US Census data and tools such as the CDC's Social Vulnerability Index (SVI), a free on-line tool (https://svi.cdc.gov/) that uses national-level data and may be modified for each state or region. The databases must be updated frequently to remain current, as people move, die, have additional children, etc. The use of new technologies such as GIS to map the vulnerabilities of populations will help planners to target at-risk regions and populations for whom climate health risks are greatest, to reduce injuries and fatalities among vulnerable populations. (Case examples of emergency planning for such populations are given in Chapter 4, Environmental Justice.)

REFERENCES

Africa Renewable Energy Access Program (AFREA). (2011). *Wood-based biomass energy development for sub-Saharan Africa*. Washington, D.C.: The World Bank.

Aiming, L. (2009). The mechanism and hazard produced by the 2008 Mw8.0 Sichuan earthquake, China. *The Almanac of China, 40*. (in Japanese)

Akachi, Y., Goodman, D., & Parker, D. (2009). *Global climate change and child health: a review of pathways, impacts and measures to improve the evidence base. Innocenti Discussion Papers 2009*. Florence, Italy: UNICEF Innocenti Research Centre.

Alessa, L., Kliskey, A., Williams, P., & Barton, M. (2008). Perception of change in freshwater in remote resource-dependent Arctic communities. *Global Environmental Change, 18*, 153-164. doi:10.1016/j.gloenvcha.2007.05.007

Arrieta, M. I., Foreman, R. D., Crook, E. D., & Icenogle, M. L. (2009). Providing continuity of care for chronic diseases in the aftermath of Katrina: From field experience to policy recommendations. *Disaster Medicine and Public Health Preparedness, 3*(03), 174–182. doi:10.1097/DMP.0b013e3181b66ae4 PMID:19865042

Balbus, J., Crimmins, A., Gamble, J.L., Easterling, D.R., Kunkel, K.E., Saha, S., & Sarofim, M.C. (2016). Introduction: Climate Change and Human Health. In *The Impacts of Climate Change on Human Health in the United States: A Scientific Assessment* (Ch. 1, pp. 25–42). U.S. Global Change Research Program, Washington, DC. doi:10.7930/J0VX0DFW

Balbus, J.M., & Malina, C. (2009). Identifying vulnerable subpopulations for climate change health effects in the United States. *Journal of Occupational and Environmental Medicine, 51*(1), 33-37. doi:10.1097/JOM.0b013e318193e12e

Becker-Blease, K.A., Turner, H.A., & Finkelhor, D. (2010). Disasters, victimization, and children's mental health. *Child Development, 81*(4), 1040-1052. doi:10.1111/j.1467-8624.2010.01453.x

Bernstein, A.S., & Myers, S.S. (2011). Climate change and children's health. *Current Opinion in Pediatrics, 23*(2), 221-226. doi:10.1097/MOP.0b013e3283444c89

Booth, S., & Zeller, D. (2005). Mercury, food webs, and marine mammals: Implications of diet and climate change for human health. *Environmental Health Perspectives, 113*(5), 521-526. doi:10.1289/ehp.7603

Brault, M. W. (2012). Americans With Disabilities: 2010. Current Population Reports (pp. 70–131). Washington, D.C.: U.S. Census Bureau. Retrieved from http://www.census.gov/prod/2012pubs/p70-131.pdf

Brunkard, J., Namulanda, G., & Ratard, R. (2008). Hurricane Katrina deaths, Louisiana, 2005. *Disaster Medicine and Public Health Preparedness, 2*(4), 215-223. doi:10.1097/DMP.0b013e31818aaf55

Bureau of Labor Statistics (BLS). (2015). Economic News Release: Table A. Employment Status of the Civilian Noninstitutional Population by Disability Status and Age, 2012 and 2013 Annual Averages. U.S. Department of Labor, Bureau of Labor Statistics. Retrieved from http://www.bls.gov/news.release/disabl.a.htm

Bush, K.F., Fossani, C.L., Li, S., Mukherjee, B., Gronlund, C.J., & O'Neill, M.S. (2014). Extreme precipitation and beach closures in the Great Lakes region: Evaluating risk among the elderly. *International Journal of Environmental Research and Public Health, 11*(2), 2014-2032. doi:10.3390/ijerph110202014

Callaghan, W.M., Rasmussen, S.A., Jamieson, D.J., Ventura, S.J., Farr, S.L., … & Posner, S.F. (2007). Health concerns of women and infants in times of natural disasters: Lessons learned from Hurricane Katrina. *Maternal and Child Health Journal, 11*(4), 307-311. doi:10.1007/s10995-007-0177-4

Carter, L. M. et al.. (2014). Southeast and the Caribbean. In J. M. Melillo, T. Richmond, & G. W. Yohe (Eds.), *Climate Change Impacts in the United States: The Third National Climate Assessment* (pp. 396–417). Washington, D.C.: U.S. Global Change Research Program.

Centers for Disease Control and Prevention (CDC). (2010). Public health workbook to define, locate, and reach special, vulnerable, and at-risk populations in an emergency. Atlanta, GA: US Department of Health and Human Services. Retrieved from http://www.cdc.gov/foodsafety/diseases/campylobacter/index.htmlhttp://www.bt.cdc.gov/workbook/pdf/ph_workbookFINAL.pdf

Centers for Disease Control and Prevention (CDC). (2011). Health Disparities and Inequalities Report-United States, 2011. *MMWR. Morbidity and Mortality Weekly Report, 60*(Suppl.), 1–116. Retrieved from http://www.cdc.gov/mmwr/pdf/other/su6001.pdf

Centers for Disease Control and Prevention (CDC). (2013). Health Disparities and Inequalities Report-United States 2013. *MMWR. Morbidity and Mortality Weekly Report, 62*(Suppl. 3), 1–187. PMID:23302815

Centers for Disease Control and Prevention (CDC). (2015a). Confirmed Lyme Disease Cases by Age and Sex--United States, 2001-2010. Atlanta, GA: Centers for Disease Control and Prevention. Retrieved from http://www.cdc.gov/lyme/stats/chartstables/incidencebyagesex. html

Centers for Disease Control and Prevention (CDC). (2015b). Disability and Health: Disability Overview. Atlanta, GA: Centers for Disease Control and Prevention. Retrieved from http://www.cdc.gov/ncbddd/disabilityandhealth/disability. html

Centers for Disease Control and Prevention (CDC). (2015c). Planning for an emergency: strategies for identifying and engaging at-risk groups. A guidance document for Emergency Managers. Atlanta. GA: US Department of Health and Human Services. Retrieved from https://www.cdc.gov/nceh/hsb/disaster/atriskguidance.pdf

Centers for Disease Control and Prevention (CDC). (2016). Mold cleanup and remediation. Atlanta, GA: Centers for Disease Control and Prevention. Retrieved from http://www.cdc.gov/mold/cleanup.htm

Centers for Disease Control and Prevention (CDC). (2017). Asthma's impact on the nation. Atlanta, GA: Centers for Disease Control and Prevention. Retrieved from https://www.cdc.gov/asthma/impacts_nation/asthmafactsheet.pdf

Chen, J., Wilkinson, D., Richardson, R. B., & Waruszynski, B. (2009). Issues, considerations, and recommendations on emergency preparedness for vulnerable population groups. *Radiation Protection Dosimetry, 134*(3-4), 132–135. doi:10.1093/rpd/ncp083 PMID:19435730

Cheruvelil, J. J., & Barton, B. (2013). *Adapting to the Effects of Climate Change on Wild Rice*. Great Lakes Lifeways Institute.

Chitra, T. V., & Panicker, S. (2011). Maternal and fetal outcome of dengue fever during pregnancy. *Journal of Vector Borne Diseases, 48*(4), 210–213. Retrieved from http://www.mrcindia.org/journal/issues/484210.pdf PMID:22297282

Cho, R. (2016). *The Damaging Effects of Black Carbon. State of the Planet. Earth Institute*. New York: Columbia University.

Choi, H., Rauh, V., Garfinkel, R., Tu, Y., & Perera, F.P. (2008). Prenatal exposure to airborne polycyclic aromatic hydrocarbons and risk of intrauterine growth restriction. *Environmental Health Perspectives, 116*(5), 658-665. doi:10.1289/ehp.10958

Costello, A., Abbas, M., Allen, A., Ball, S., Bell, S., Bellamy, R., & Patterson, C. et al. (2009). Managing the Health effects of climate change: Lancet and University College London Institute for Global Health Commission. *Lancet, 373*(9676), 1693–1733. doi:10.1016/S0140-6736(09)60935-1 PMID:19447250

Cozzetto, K., Chief, K. Dittmer, K., Brubaker, M., Gough, R., Souza, K., Ettawageshik, F., Wotkyns, S., Opitz-Stapleton, S., Duren, S., & Chavan, P. (2013). Climate change impacts on the water resources of American Indians and Alaska Natives in the U.S. *Climatic Change, 120*(3), 569-584. doi: 10.1007/s10584-013- 0852-y

CSDH. (2008). Closing the Gap in a Generation: Health Equity through Action on the Social Determinants of Health. Final Report of the Commission on Social Determinants of Health. World Health Organization, Geneva. Retrieved from http://www.who.int/social_determinants/final_report/csdh_finalreport_2008.pdf

Cunsolo Willox, A., (2012). Climate change as the work of mourning. *Ethics & the Environment, 17*(2), 137-164. doi: 10.2979/ethicsenviro.17.2.137

Cunsolo Willox, A., Harper, S.L., Edge, V.L., Landman, K., Houle, K., Ford, J.D., & Rigolet Inuit Community Government (2013). The land enriches the soul: On climatic and environmental change, affect, and emotional health and well-being in Rigolet, Nunatsiavut, Canada. *Emotion, Space and Society, 6*, 14-24. doi:10.1016/j.emospa.2011.08.005

Cunsolo Willox, A., Stephenson, E., Allen, J., Bourque, F., Drossos, A., Elgarøy, S. ... Wexler, L. (2015). Examining relationships between climate change and mental health in the Circumpolar North. *Regional Environmental Change, 15*(1), 169-182. doi: 10.1007/s10113-014-0630-z

Dechet, A.M., Yu, P.A., Koram, N., & Painter, J. (2008). Nonfoodborne Vibrio infections: An important cause of morbidity and mortality in the United States, 1997–2006. *Clinical Infectious Diseases, 46*(7), 970-976. doi:10.1086/529148

DHHS. (2014). National Healthcare Disparities Report 2013. AHRQ Publication No. 14-0006. Rockville, MD: U.S. Department of Health and Human Services, Agency for Healthcare Research and Quality. Retrieved from http://www.ahrq.gov/research/findings/nhqrdr/nhdr13/index.html

Dijkstra, J.A., Buckman, K.L., Ward, D., Evans, D.W., Dionne, M., & Chen, C.Y. (2013). Experimental and natural warming elevates mercury concentrations in estuarine fish. *PLoS ONE, 8*, e58401. doi:.pone.005840110.1371/journal

Donner, W., & Rodríguez, H. (2008). Population composition, migration and inequality: The influence of demographic changes on disaster risk and vulnerability. *Social Forces, 87*(2), 1089–1114. doi:10.1353/sof.0.0141

Doyle, J.T., Redsteer, M.H., & Eggers, M.J. (2013). Exploring effects of climate change on Northern Plains American Indian health. *Climatic Change, 120*(3), 643-655. doi.org/ 10.1007/s10584-013-0799-z

Drayna, P., McLellan, S.L., Simpson, P., Li, S.-H., & Gorelick, M.H. (2010). Association between rainfall and pediatric emergency department visits for acute gastrointestinal illness. *Environmental Health Perspectives, 118*(10), 1439-1443. doi:10.1289/ehp.0901671

Dybas, C. L. (2009). Minnesotas moose: Ghosts of the northern forest? *Bioscience*, *59*(10), 824–828. doi:10.1525/bio.2009.59.10.3

Ebi, K. L., Sussman, F. G., & Wilbanks, T. J. (2008). Analyses of the Effects of Global Change on Human Health and Welfare and Human Systems. A Report by the U.S. Climate Change Science Program and the Subcommittee on Global Change Research (J. L. Gamble, Ed.). Washington, D.C.: U.S. Environmental Protection Agency. Retrieved from http://downloads.globalchange.gov/sap/sap4-6/sap4-6-final-report-all.pdf

Eisenman, D. P., Cordasco, K. M., Asch, S., Golden, J. F., & Glik, D. (2007). Disaster planning and risk communication with vulnerable communities: Lessons from Hurricane Katrina. *American Journal of Public Health*, *97*(Suppl. 1), S109–S115. doi:10.2105/AJPH.2005.084335 PMID:17413069

Eneriz-Wiemer, M., Sanders, L. M., Barr, D. A., & Mendoza, F. S. (2014). Parental limited English proficiency and health outcomes for children with special health care needs: A systematic review. *Academic Pediatrics*, *14*(2), 128–136. doi:10.1016/j.acap.2013.10.003 PMID:24602575

Erwin, P.C., Jones, T.F., Gerhardt, R.R., Halford, S.K., Smith, A.B., Patterson, L.E.R., Gottfried, K.L., Burkhalter, K.L., Nasci, R.S., & Schaffner, W. (2002). La Crosse encephalitis in eastern Tennessee: Clinical, environmental, and entomological characteristics from a blinded cohort study. *American Journal of Epidemiology, 155*(11), 1060-1065. doi:10.1093/aje/155.11.1060

Evans-Campbell, T. (2008). Historical trauma in American Indian/Native Alaska communities: A multilevel framework for exploring impacts on individuals, families, and communities. *Journal of Interpersonal Violence*, *23*(3), 316–338. doi:10.1177/0886260507312290 PMID:18245571

Fairbank, J. A., Putnam, F. W., & Harris, W. W. (2014). Child traumatic stress: Prevalence, trends, risk, and impact. In M. J. Friedman, T. M. Keane, & P. A. Resick (Eds.), *Handbook of PTSD: Science and Practice* (2nd ed., pp. 121–145). New York: Guilford Press.

Filiberto, D., Wethington, E., Pillemer, K., Wells, N., Wysocki, M., & Parise, J. T. (2010). Older people and climate change: Vulnerability and health effects. *Generations, 33*(4), 19-25. Retrieved from http://www.asaging.org/blog/older-people-and-climate-change-vulnerability-and-health-effects

Fischer, G., Shah, M., Tubiello, F. N., & van Velhuizen, H. (2005). Socio-economic and climate change impact on agriculture: an integrated assessment 1990–2080. *Philosophical Transactions of the Royal Society of London. Series B, Biological Sciences*, *360*(1463), 2067–2083. doi:10.1098/rstb.2005.1744 PMID:16433094

Flanagan, B.E., Gregory, E.W., Hallisey, E.J., Heitgard, J.L., & Lewis, B. (2011). A social vulnerability index for disaster management. *J. Homeland Secur. Emerg. Manage., 8*(1).

Fox, M.H., White, G.W., Rooney, C., & Cahill, A. (2010). The psychosocial impact of Hurricane Katrina on persons with disabilities and independent living center staff living on the American Gulf Coast. *Rehabilitation Psychology, 55*(3), 231-240. doi:10.1037/a0020321

Frey, W. (2011). The New Metro Minority Map: Regional Shifts in Hispanics, Asians, and Blacks from Census 2010. Brookings Institution, Washington, D.C. Retrieved from http://www.brookings.edu/papers/2011/0831_census_race_frey.aspx

Friel, S., Marmot, M., McMichael, A.J., Kjellstrom, T., & Vågerö, D. (2008). Global health equity and climate stabilisation: A common agenda. *The Lancet, 372*(9650), 1677-1683. doi:10.1016/s0140-6736(08)61692-x

Fuentes-Afflick, E., & Hessol, N. A. (2009). Immigration status and use of health services among Latina women in the San Francisco Bay Area. *Journal of Women's Health*, *18*(8), 12751280. doi:10.1089/jwh.2008.1241 PMID:19627243

Gamble, J. L., Balbus, J., Berger, M., Bouye, K., & Campbell, V. …Wolkin, A.F. (2016). Ch. 9: Populations of Concern. IN: *The Impacts of Climate Change on Human Health in the United States: A Scientific Assessment.* Washington, D.C.: U.S. Global Change Research Program. Retrieved from https://health2016.globalchange.gov/

Gilchrist, J., Haileyesus, T., Murphy, M., Comstock, R. D., Collins, C., McIlvain, N., & Yard, E. (2010). Heat illness among high school athletes --- United States, 2005-2009. *MMWR. Morbidity and Mortality Weekly Report, 59*(32), 1009–1013. Retrieved from http://www.cdc.gov/mmwr/preview/mmwrhtml/mm5932a1.htm PMID:20724966

Gottschalk, A.W. & Andrish, J.T. (2011). Epidemiology of sports injury in pediatric athletes. *Sports Medicine and Arthroscopy Review, 19*(1), 2-6. doi:10.1097/ JSA.0b013e31820b95fc

Hamilton, B. E., Sutton, P. D., Mathews, T. J., Martin, J. A., & Ventura, S. J. (2009). The effect of Hurricane Katrina: Births in the U.S. Gulf Coast region, before and after the storm. *National Vital Statistics Reports*, *58*(2), 1–28, 32. Retrieved from http://www.cdc.gov/nchs/data/nvsr/nvsr58/nvsr58_02.pdf PMID:19754006

Harville, E.W., Xiong, X., & Buekens, P. (2009). Hurricane Katrina and perinatal health. *Birth, 36*(4), 325-331. doi:10.1111/j.1523-536X.2009.00360.x

Hausfater, P., Megarbane, B., Dautheville, S., Patzak, A., Andronikof, M., … & Riou, B. (2009). Prognostic factors in non-exertional heatstroke. *Intensive Care Medicine, 36*(2), 272-280. doi:10.1007/ s00134-009-1694-y

Hilborn, E. D., Roberts, V. A., Backer, L., DeConno, E., Egan, J. S., & Hlavsa, M. C. et al. (2014). Algal bloom–associated disease outbreaks among users of freshwater lakes — United States, 2009–2010. *MMWR. Morbidity and Mortality Weekly Report*, *63*(1), 11–15. Retrieved from http://www.cdc.gov/mmwr/preview/mmwrhtml/mm6301a3.htm PMID:24402467

Hodge, D.R. & Limb, G.E. (2010). A Native American perspective on spiritual assessment: The strengths and limitations of a complementary set of assessment tools. *Health & Social Work, 35*(2), 121-131. doi:10.1093/hsw/35.2.121

Hodge, D.R., Limb, G.E., & Cross, T.L. (2009). Moving from colonization toward balance and harmony: A Native American perspective on wellness. *Social Work, 54*(3), 211-219. doi:10.1093/sw/54.3.211

Homer, C. S. E., Hanna, E., & McMichael, A. J. (2009). Climate change threatens the achievement of the millennium development goal for maternal health. *Midwifery*, *25*(6), 606–612. doi:10.1016/j.midw.2009.09.003 PMID:19836867

Hoover, E., Cook, K., Plain, R., Sanchez, K., Waghiyi, V., Miller, P., & Carpenter, D. O. et al. (2012). Indigenous peoples of North America: Environmental exposures and reproductive justice. *Environmental Health Perspectives*, *120*, 1645–1649. PMID:22899635

Humes, K. R., Jones, N. A., & Ramirez, R. R. (2011). Overview of Race and Hispanic Origin: 2010. 2010 Census Briefs C2010BR-02. U.S. Census Bureau. Retrieved from http://www.census.gov/prod/cen2010/briefs/c2010br-02.pdf

IPCC. (2014). Summary for policymakers. In C.B. Field, V.R. Barros, D.J. Dokken et al. (Eds.), *Climate Change 2014: Impacts, Adaptation, and Vulnerability. Part A: Global and Sectoral Aspects. Contribution of Working Group II to the Fifth Assessment Report of the Intergovernmental Panel on Climate Change*. Cambridge, UK: Cambridge University Press. Retrieved from http://ipcc.ch/pdf/assessment-report/ar5/wg2/ar5_wgII_spm_en.pdf

Jagai, J.S., Griffiths, J.K., Kirshen, P.K., Webb, P., & Naumova, E.N. (2012). Seasonal patterns of gastrointestinal illness and streamflow along the Ohio River. *International Journal of Environmental Research and Public Health, 9*(5), 1771-1790. doi:10.3390/ijerph9051771

Janakiraman, V. (2008). Listeriosis in pregnancy: Diagnosis, treatment, and prevention. *Reviews in Obstetrics and Gynecology*, *1*(4), 179–185. PMID:19173022

Jayachandran, S. (2009). Air quality and early-life mortality: Evidence from Indonesia's wildfires. *The Journal of Human Resources*, *44*(4), 916–954. doi:10.3368/jhr.44.4.916

Joseph, N. T., Matthews, K. A., & Myers, H. F. (2014). Conceptualizing health consequences of Hurricane Katrina from the perspective of socioeconomic status decline. *Health Psychology*, *33*(2), 139–146. doi:10.1037/a0031661 PMID:23527519

Keim, M. E. (2008). Building human resilience: The role of public health preparedness and response as an adaptation to climate change. *American Journal of Preventive Medicine*, *35*(5), 508–516. doi:10.1016/j.amepre.2008.08.022 PMID:18929977

Kellogg, J., Wang, J., Flint, C., Ribnicky, D., Kuhn, P., De Mejia, E. G. I., & Lila, M. A. et al. (2010). Alaskan wild berry resources and human health under the cloud of climate change. *Journal of Agricultural and Food Chemistry, 58*(7), 38843900. doi:10.1021/jf902693r PMID:20025229

Kerr, Z.Y., Casa, D.J., Marshall, S.W., & Comstock, R.D. (2013). Epidemiology of exertional heat illness among U.S. high school athletes. *American Journal of Preventive Medicine, 44*(1), 8-14. doi:10.1016/j.amepre.2012.09.058

Klinenberg, E. (2002). Dying alone: an interview with Eric Klinenberg. Chicago: University of Chicago Press. Retrieved from http://www.press.uchicago.edu/Misc/Chicago/443213in.html

Klinenberg, E. (2015). *Heat Wave: a Social Autopsy of Disaster in Chicago* (2nd ed.). Chicago: University of Chicago Press.

Knowlton, K., Rotkin-Ellman, M., King, M. G., Margolis, H. G., Smith, D., Solomon, G., & English, P. et al. (2009). The 2006 California heat wave: Impacts on hospitalizations and emergency department visits. *Environmental Health Perspectives, 117*(1), 61–67. doi:10.1289/ehp.11594 PMID:19165388

Laditka, S.B., Laditka, J.N., Xirasagar, S., Cornman, C.B., Davis, C.B., & Richter, J.V.E. (2008). Providing shelter to nursing home evacuees in disasters: Lessons from Hurricane Katrina. *American Journal of Public Health, 98*(7), 1288-1293. doi:10.2105/ajph.2006.107748

Lane, K., Charles-Guzman, K., Wheeler, K., Abid, Z., Graber, N., & Matte, T. (2013). Health effects of coastal storms and flooding in urban areas: A review and vulnerability assessment. *Journal of Environmental and Public Health*. doi:10.1155/2013/913064 PMID:23818911

Lenarz, M. S., Nelson, M. E., Schrage, M. W., & Edwards, A. J. (2009). Temperature mediated moose survival in northeastern Minnesota. *The Journal of Wildlife Management, 73*(4), 503–510. doi:10.2193/2008-265

Lin, S., Luo, M., Walker, R.J., Liu, X., Hwang, S.-A., & Chinery, R. (2009). Extreme high temperatures and hospital admissions for respiratory and cardiovascular diseases. *Epidemiology, 20*(5), 738-746. doi:10.1097/EDE.0b013e3181ad5522

Lindsey, N. P., Staples, J. E., Lehman, J. A., & Fischer, M. (2010). Surveillance for human West Nile virus disease--United States, 1999-2008. *MMWR. Surveillance Summaries, 59*(SS02), 1–17. Retrieved from http://www.cdc.gov/mmwr/preview/mmwrhtml/ss5902a1.htm PMID:20360671

Little, B., & Gill, J. Schulte, J., Young, S., Horton, J., … & Bayleyegn, T. (2004). Rapid assessment of the needs and health status of older adults after Hurricane Charley— Charlotte, DeSoto, and Hardee Counties, Florida, August 27-31, 2004. *MMWR. Morbidity and Mortality Weekly Report, 53*(36), 837-840. Retrieved from http:// www.cdc.gov/mmwr/preview/mmwrhtml/mm5336a2.htm

Luber, G., Knowlton, J., Balbus, J., Frumkin, H., Hayden, M., Hess, J., ... (2014). Human Health Climate Change Impacts in the United States. In J.M. Melillo, T.C. Richmond, & G.H. Yohe (Eds.), *The Third National Climate Assessment* (Ch. 9, p. 222). U.S. Global Climate Change Research Program. Retrieved from http://nca2014.globalchange.gov/report/sectors/human-health

Luber, G., & McGeehin, M. (2008). Climate change and extreme heat events. *American Journal of Preventive Medicine, 35*(5), 429-435. doi:10.1016/j.amepre.2008.08.021

Lynn, K., Daigle, J. Hoffman, J., Lake, F., Michelle, N., Ranco, D., Viles, C., Voggesser, G., & Williams, P. (2013). The impacts of climate change on tribal traditional foods. *Climatic Change, 120*(3), 545-556. doi:10.1007/s10584-013-0736-1

Makri, A., & Stilianakis, N.I. (2008). Vulnerability to air pollution health effects. *International Journal of Hygiene and Environmental Health, 211*(3), 326-336. doi:10.1016/j.ijheh.2007.06.005

Maldonado, C. Z., Rodriguez, R. M., Torres, J. R., Flores, Y. S., & Lovato, L. M. (2013). Fear of discovery among Latino immigrants presenting to the emergency department. *Academic Emergency Medicine, 20*(2), 155–161. doi:10.1111/acem.12079 PMID:23406074

Manangan, A. P., Uejio, C. K., Saha, S., Schramm, P. J., Marinucci, G. D., . . . Luber, G. (2014). Assessing Health Vulnerability to Climate Change: A Guide for Health Departments. Climate and Health Technical Report Series, Centers for Disease Control and Prevention, Atlanta, GA. Retrieved from http://www.cdc.gov/climateandhealth/ pubs/AssessingHealthVulnerabilitytoClimateChange.pdf

Martin-Latry, K., Goumy, M.P., Latry, P., Gabinski, C., Bégaud, B., Faure, I., & Verdoux, H. (2007). Psychotropic drugs use and risk of heat-related hospitalisation. *European Psychiatry, 22*(6), 335-338. doi:10.1016/j.eurpsy.2007.03.007

McMichael, A. J. (2013). Globalization, climate change, and human health. *The New England Journal of Medicine, 368*(14), 1335–1343. doi:10.1056/NEJMra1109341 PMID:23550671

MedNet. (2010). Impact of maternal nutrition on fetal development. Medical Education Network. Retrieved from http://www.mednet.ca/en/report/impact-of-maternal-nutrition-on-fetal-developmen.html

Melillo, J. M., Richmond, T. C., & Yohe, G. W. (Eds.). (2014). *Climate Change Impacts in the United States: The Third National Climate Assessment. U.S. Global Change Research Program.* doi:10.7930/J0Z31WJ2

Merriam-Webster Dictionary. (2017). Retrieved from https://www.merriam-webster.com/thesaurus/vulnerability

Milburn, M. P. (2004). Indigenous nutrition: Using traditional food knowledge to solve contemporary health problems. *American Indian Quarterly, 28*(3), 411–434. doi:10.1353/aiq.2004.0104

Miranda, M.L., Hastings, D.A., Aldy, J.E., & Schlesinger, W.H. (2011). The environmental justice dimensions of climate change. *Environmental Justice, 4*(1), 17-25. doi:10.1089/env.2009.004628

Nakashima, D. J., Galloway, K., McLean, H. D., Thulstrup, A., Castillo, R., & Rubis, J. T. (2012). Weathering Uncertainty: Traditional Knowledge for Climate Change Assessment and Adaptation. UNESCO, Paris and UNU, Darwin.

National Center for Health Statistics (NCHS). (2014). Table 49. Disability measures among adults aged 18 and over, by selected characteristics: United States, selected years 1997– 2012. In *Health, United States, 2013: With Special Feature on Prescription Drugs. National Center for Health Statistics, Hyattsville, MD* (pp. 172-173). Retrieved from http://www. cdc.gov/nchs/data/hus/hus13.pdf

National Resource Council (NRC). (2012). *Disaster Resilience: A National Imperative*. National Academies Press.

Naumova, E.N., Egorov, A.I., Morris, R.D., & Griffiths, J.K. (2003). The elderly and waterborne Cryptosporidium infection: Gastroenteritis hospitalizations before and during the 1993 Milwaukee outbreak. *Emerging Infectious Diseases, 9*(4), 418-425. doi:10.3201/eid0904.020260

Nelson, N.G., Collins, C.L., Comstock, R.D., & McKenzie, L.B. (2011). Exertional heat-related injuries treated in emergency departments in the U.S., 1997–2006. *Amer. J. Preventive Med., 40*(1), 54-60. doi:10.1016/j.amepre.2010.09.031

Nordon, C., Martin-Latry, K., Roquefeuil, L.D., Latry, P., Bégaud, B., … Verdoux, H. (2009). Risk of death related to psychotropic drug use in older people during the European 2003 heatwave: A population-based case–control study. *Amer. J. Geriatric Psychiatry, 17*(12), 1059-1067. doi:10.1097/JGP.0b013e3181b7ef6e

O'Neill, M.S. & Ebi, K.L. (2009). Temperature extremes and health: Impacts of climate variability and change in the United States. *J. Occupational and Environmental Medicine, 51*(1), 13-25. doi:10.1097/JOM.0b013e318173e122

Ortega, A. N., Fang, H., Perez, V. H., Rizzo, J. A., Carter-Pokras, O., Wallace, S. P., & Gelberg, L. (2007). Health care access, use of services, and experiences among undocumented Mexicans and other Latinos. *Archives of Internal Medicine*, *167*(21), 2354–2360. doi:10.1001/archinte.167.21.2354 PMID:18039995

Ostro, B., Roth, L., Malig, B., & Marty, M. (2009). The effects of fine particle components on respiratory hospital admissions in children. *Environmental Health Perspectives, 117*(3), 475-480. doi:10.1289/ehp.11848

Otani, Y., Ando, T., Atobe, K., Hauden, A., Kao, S.-Y., Saito, K., & Fukunaga, K. et al. (2012). Comparison of Two Large Earthquakes: The 2008 Sichuan Earthquake and the 2011 East Japan Earthquake. *The Keio Journal of Medicine*, *61*(1), 35–39. doi:10.2302/kjm.61.35 PMID:22410538

Parker, J.D., L.J. Akinbami, and T.J. Woodruff, (2009). Air pollution and childhood respiratory allergies in the United States. *Environmental Health Perspectives, 117*(1), 140-147. doi:10.1289/ehp.11497

Polin, R. A., & Abman, S. H. (2011). *Thermoregulation. In Fetal and Neonatal Physiology* (4th ed., pp. 615–670). Philadelphia, PA: Elsevier.

Ranada, P. (2016). PH drops in 2016 list of countries vulnerable to climate change. *Rappler.com*. Retrieved from http://www.rappler.com/science-nature/environment/113064-philippines-2016-climate-change-vulnerability-index

Rasmussen, S.A., Jamieson, D.J., & Uyeki, T.M. (2012). Effects of influenza on pregnant women and infants. *American Journal of Obstetrics and Gynecology, 207*(3), S3-S8. doi:10.1016/j.ajog.2012.06.068

Redsteer, M. H., Kelley, K. B., Francis, H., & Block, D. (2011). Disaster Risk Assessment Case Study. UN Office for Disaster Risk Reduction & US Geological Survey, Reston, VA. Retrieved from http://www.preventionweb.net/english/hyogo/gar/2011/en/bgdocs/Redsteer_Kelley_Francis_&_Block_2010.pdf

Regalado, J., Pérez-Padilla, R., Sansores, R., Páramo Ramirez, J. I., Brauer, M., Paré, P., & Vedal, S.Instituto Nacional de Enfermedades Respiratorias. (2006). The effect of biomass burning on respiratory symptoms and lung function in rural Mexican women. *American Journal of Respiratory and Critical Care Medicine, 174*(8), 901–905. doi:10.1164/rccm.200503-479OC PMID:16799080

Ritz, B., Wilhelm, M., Hoggatt, K.J., & Ghosh, J.K.C. (2007). Ambient air pollution and preterm birth in the Environment and Pregnancy Outcomes Study at the University of California, Los Angeles. *Amer J Epidem, 166*(9), 1045-1052. doi:10.1093/aje/kwm181

Roberts, J.W., Wallace, L.A., Camann, D.E., Dickey, P., Gilbert, S.G., Lewis, R.G., & Takaro, T.K. (2009). Monitoring and reducing exposure of infants to pollutants in house dust. *Reviews of Environmental Contamination and Toxicology, 201*, 1-39. doi:10.1007/978-1-4419-0032-6_1

Rylander, C., Odland, J. O., & Sandanger, T. M. (2013). *Climate change and the potential effects on maternal and pregnancy outcomes: An assessment of the most vulnerable--the mother, fetus, and newborn child. Global Health Action, 6*. doi:10.3402/gha.v6i0.19538

Sarche, M., & Spicer, P. (2008). Poverty and health disparities for American Indian and Alaska Native children. *Annals of the New York Academy of Sciences, 1136*(1), 126–136. doi:10.1196/annals.1425.017 PMID:18579879

Satia, J. A. (2009). Diet-related disparities: Understanding the problem and accelerating solutions. *Journal of the American Dietetic Association, 109*(4), 610–615. doi:10.1016/j.jada.2008.12.019 PMID:19328255

Schulte, P.A., & Chun, H. (2009). Climate change and occupational safety and health: Establishing a preliminary framework. *J Occupational and Environmental Hygiene, 6*(9), 542-554. doi:10.1080/15459620903066008

Semenza, J. C., McCullough, J. E., Flanders, W. D., McGeehin, M. A., & Lumpkin, J. R. (1999). Excess hospital admissions during the July 1995 heat wave in Chicago. *American Journal of Preventive Medicine, 16*(4), 269–277. doi:10.1016/S0749-3797(99)00025-2 PMID:10493281

Shannon, M.W., Best, D., Binns, H.J., Forman, J.A., Johnson, C.L., Karr, C.J., Kim, J.J., Mazur, L.J., Roberts, J.R., & Shea, K.M. (2007). Global climate change and children's health. *Pediatrics, 120*(5), 1149-1152. doi:10.1542/peds.2007-2645

Sheffield, P.E., & Landrigan, P.J. (2011). Global climate change and children's health: Threats and strategies for prevention. *Environmental Health Perspectives, 119*(3), 291-298. doi:10.1289/ehp.1002233

Stöllberger, C., Lutz, W., & Finsterer, J. (2009). Heat-related side-effects of neurological and non-neurological medication may increase heatwave fatalities. *European J. Neur., 16*(7), 879-882. doi:10.1111/j.14681331.2009.02581.x

Strickland, M.J., Darrow, L.A., Klein, M., Flanders, W.D., Sarnat, J.A., Waller, L.A., Sarnat, S.E., Mulholland, J.A., & Tolbert, P.E. (2010). Short-term associations between ambient air pollutants and pediatric asthma emergency department visits. *American Journal of Respiratory and Critical Care Medicine, 182*(3), 307-316. doi:10.1164/rccm.200908-1201OC

Sue Kirkman, , M.S., Briscoe, V.J., Clark, N., Florez, H., Haas, L.B., . . . Swift, C.S. (2012, December). Diabetes in older adults. *Diabetes Care, 35*(12), 2650–2664. doi:10.2337/dc12-1801 PMID:23100048

Tees, M.T., Harville, E.W., Xiong, X., Buekens, P., Pridjian, G., & Elkind-Hirsch, K. (2010). Hurricane Katrina-related maternal stress, maternal mental health, and early infant temperament. *Maternal and Child Health Journal, 14*(4), 511518. doi:10.1007/s10995-009-0486-x

Tribal Water Working Group (TWWG). (2012). Water in Indian Country: Challenges and Opportunities. Tribal Water Working Group. Retrieved from http://uttoncenter.unm.edu/pdfs/2012White_Paper.pdf

Turner, B. L., Kasperson, R. E., Matson, P. A., McCarthy, J. J., Corell, R. W., . . . Schiller, A. (2003). A framework for vulnerability analysis in sustainability science. *Proceedings of the National Academy of Sciences, 100*(14), 8074-8079. doi:10.1073/pnas.1231335100

U.S. Census Bureau. (2008). An Older and More Diverse Nation by Midcentury. Retrieved from https://www.census.gov/newsroom/releases/archives/population/cb08-123.html

U.S. Census Bureau. (2010). The next four decades, the older population in the United States, 2010-2050. Population estimate and projections. U.S. Department of Commerce, Economics and Statistics Division. Washington, D.C.: U.S. Census Bureau

U.S. Census Bureau. (2014a). Annual Estimates of the Resident Population by Sex, Race, and Hispanic Origin for the United States, States, and Counties: April 1, 2010 to July 1, 2013. U.S. Census Bureau American Fact Finder. Retrieved from http://factfinder.census.gov/faces/tableservices/jsf/pages/productview.xhtml?src=bkmk

U.S. Census Bureau. (2014b). *2014 National Population Projections: Summary Tables. Table 9. Projections of the Population by Sex and Age for the United States: 2015 to 2060*. U.S. Department of Commerce.

U.S. Environmental Protection Agency (EPA). (2016). Climate change and the health of occupational groups. Environmental justice. Washington, D.C.: US EPA.

U.S. Environmental Protection Agency (EPA). (2017). Environmental justice. Washington, D.C.: US EPA. Retrieved from https://www.epa.gov/environmentaljustice

U.S. EPA, Indian Health Service, Department of Agriculture, and Department of Housing and Urban Development. (2008). Meeting the Access Goal: Strategies for Increasing Access to Safe Drinking Water and Wastewater Treatment to American Indian and Alaska Native Homes. Infrastructure Task Force Access Subgroup. Retrieved from http://www.epa.gov/tp/pdf/infra-tribal-access-plan.pdf

USGCRP. (2016). *Impact of Climate Change on Human Health in the United States: A Scientific Assessment* (A. J. Crimmins et al., Eds.). Washington, D.C.: U.S. Global Change Research Program.

van Zutphen, A.R., Lin, S., Fletcher, B.A., & Hwang, S.A. (2012). A population-based case- control study of extreme summer temperature and birth defects. *Environmental Health Perspectives, 120*(10), 1443-1449. doi:10.1289/ehp.1104671

Vargas Bustamante, A., Fang, H., Garza, J., Carter-Pokras, O., Wallace, S. P., Rizzo, J. A., & Ortega, A. N. (2010). Variations in healthcare access and utilization among Mexican immigrants: The role of documentation status. *Journal of Immigrant and Minority Health*, *14*(1), 146–155. doi:10.1007/s10903-010-9406-9 PMID:20972853

Voggesser, G., Lynn, K., Daigle, J., Lake, F.K., & Ranco, D. (2013). Cultural impacts to tribes from climate change influences on forests. *Climatic Change, 120*(3), 615-626. doi: 10.1007/s10584-013-0733-4

Wang, L., Green, F. H. Y., Smiley-Jewell, S. M., & Pinkerton, K. E. (2010). Susceptibility of the aging lung to environmental injury. *Seminars in Respiratory and Critical Care Medicine*, *31*(05), 539–553. doi:10.1055/s-0030-1265895 PMID:20941655

Wong, C. A., Gachupin, F. C., Holman, R. C., MacDorman, M. F., Cheek, J. E., Holve, S., & Singleton, R. J. (2014). American Indian and Alaska Native infant and pediatric mortality, United States, 1999–2009. *American Journal of Public Health*, *104*(Suppl. 3), S320–S328. doi:10.2105/AJPH.2013.301598 PMID:24754619

World Health Organization (WHO). (2011). *World Report on Disability*. Geneva, Switzerland: WHO.

World Health Organization (WHO). (2016). Malaria deaths in children under five. Geneva, Switzerland: WHO. Retrieved from http://www.who.int/malaria/areas/high_risk_groups/children/en/

Xu, Z., Etzel, R.A., Su, H., Huang, C., Guo, Y., & Tong, S. (2012). Impact of ambient temperature on children's health: A systematic review. *Environmental Research, 117*, 120-131. doi:10.1016/j.envres.2012.07.002

Xu, Z., Sheffield, P. E., Hu, W., Su, H., Yu, W., Qi, X., & Tong, S. (2012). *Climate change and children's health—A call for research on what works to protect children. Intern. J. Environ. Res. Public Health, 9(9)*. doi:10.3390/ijerph9093298

Younger, M., Morrow-Almeida, H. R., Vindigni, S. M., & Dannenberg, A. L. (2008). The built environment, climate change, and health: Opportunities for co-benefits. *American Journal of Preventive Medicine*, *35*(5), 517–526. doi:10.1016/j.amepre.2008.08.017 PMID:18929978

Zhou, H., Shi, L., Yuping, M., Tang, J., & Zeng, Y. (2012). *Diffusion of new technology, health services and information after a crisis: a focus group study of the Sichuan "5.12" Earthquake*. Beijing, China: John Wiley & Sons Ltd.

Zoraster, R. M. (2010). Vulnerable populations: Hurricane Katrina as a case study. *Prehospital and Disaster Medicine*, *25*(01), 74–78. doi:10.1017/S1049023X00007718 PMID:20405467

KEY TERMS AND DEFINITIONS

Adaptive Capacity: The ability of communities, institutions, or people to adjust to potential hazards, to take advantage of opportunities, or to respond to consequences. A related term, *resilience*, is the ability to prepare and plan for, absorb, recover from, and more successfully adapt to adverse events.

Asthma: A chronic disease in which a person's airways become inflamed, narrow, and swell, producing extra mucous which makes it difficult to breathe; asthma symptoms can be controlled through trigger avoidance, medication and monitoring by a healthcare provider.

Dengue Fever: A mosquito-borne viral infection that causes a flu-like illness, and occasionally develops into a potentially lethal complication called severe dengue, which is a leading cause of serious illness and death among children in some Asian and Latin American countries. The global incidence of dengue has grown dramatically; currently about half of the world's population is at risk.

Disability: Any impairment of the body or mind that limits a person's functional ability to perform certain activities or participate in life activities such as school, work, and recreation.

Exposure: Contact between a person and one or more biological, psychosocial, chemical, or physical stressors, including stressors affected by climate change, which may occur in a single instance or repeatedly over time, and in one location or over a wider geographic area.

Geographic Information System (GIS): A computer system for capturing, storing, checking, and displaying data related to positions on Earth's surface, GIS can show many different kinds of data of data on a map, which facilitates data analysis.

Influenza: An acute viral infection of the respiratory tract that has a worldwide death toll of 500 million people per year; it can be transmitted by contact with infected individuals and contaminated objects (fomites), and by inhalation of virus-laden aerosols.

Listeriosis: A serious infection caused by eating food contaminated with the bacterium *Listeria monocytogenes*, this infection is most likely to sicken pregnant women and their newborns, adults aged 65 or older, and people with weakened immune systems. In 2010, listeriosis resulted in 23,150 illnesses and 5463 deaths worldwide.

Lyme Disease: A bacterial infection transmitted primarily by *Ixodes* ticks, also known as deer ticks, and on the West Coast, black-legged ticks. Approximately 300,000 people are diagnosed with Lyme disease in the US every year.

Malaria: A life-threatening blood disease caused by parasites transmitted to humans through the bite of the *Anopheles* mosquito; in 2015, there were roughly 212 million malaria cases and an estimated 429 000 malaria deaths worldwide.

Maternal Health: The health of women during pregnancy, childbirth, and the postpartum period.

Risk: The probability of a hazardous event occurring multiplied by the expected severity of the impact of that event.

Social Vulnerability Index: A free online tool (https://svi.cdc.gov) developed by the Centers for Disease Control and Prevention to measure a community's resilience in the face of natural or human-caused disasters or disease outbreaks.

Stressors: Events or trends, whether related to climate change or other factors, that increase vulnerability to health effects.

Susceptibility to Harm, or Sensitivity: Degree to which people or communities are affected, either adversely or beneficially, by climate variability or change.

Volatile Organic Compounds (VOCs): Chemicals emitted as gases from certain solids or liquids which may have short- and long-term adverse health effects, especially indoors, where VOC concentrations are up to 10 times higher than outdoors.

Vulnerability: The tendency or predisposition to be adversely affected by climate-related health effects, and comprising three elements: exposure, sensitivity or susceptibility to harm, and the capacity to adapt to or to cope with change (see exposure, susceptibility to harm, or sensitivity, adaptive capacity).

Chapter 4
Cultivating Environmental Justice

ABSTRACT

The environmental justice movement grew out of the civil rights movement, and its aim was to provide all people with equal environmental protection. In the 1970s it became clear that African American and Hispanic children had much greater exposure to lead paint than did other children, and that hazardous waste dumps were disproportionately placed in communities of color. In 1991, the First National People of Color Environmental Leadership Summit in Washington, D.C., laid out the 17 principles of environmental justice. One of the Summit leaders was Hazel Johnson, an African American mother from Chicago who formed a nonprofit organization to clean up toxins in her neighborhood, which had the highest concentration of hazardous waste dumps in the nation. Mrs. Johnston's long battle with big industrial polluters is the focus of one of this chapter's case examples of how communities can empower their residents to fight for and achieve environmental justice.

INTRODUCTION

Climate change does not impact populations uniformly. As was discussed in Chapter 3, some populations are more at risk for adverse impacts than others, due to their age, gender, health status, race, income, origins, or occupation. Typically, when extreme weather events strike, people living in poverty, especially the very young and the very old, are hardest hit and are more likely

DOI: 10.4018/978-1-5225-3414-3.ch004

to be displaced and relocated than are other populations. They experience higher rates of injury, illness, and death than do other populations.

WHAT IS ENVIRONMENTAL JUSTICE?

The concept of environmental justice is based on the belief that all people–regardless of their race, color, national origin or income–should enjoy equally high levels of environmental protection. Low-income and minority communities are most vulnerable to environmental justice issues. These communities often have a disproportionate share of polluting facilities, which exposes their residents to greater health problems from environmental pollution (Maryland Department of the Environment, 2017). These communities may not have an organized group that can serve as a point of contact. It is important to make these communities aware of the environmental issues in their area, to empower them to participate in the policy-making process, and to enable them to access available resources in order to make their communities safe, healthy and sustainable (Maryland Department of the Environment, 2017).

History and Principles of Environmental Justice

The environmental justice (EJ) movement emerged in the 1980s in the United States. It grew out of the Civil Rights Movement, beginning with the passage of the Civil Rights Act of 1964, which prohibited discrimination based on race, color, or national origin (Maryland Department of the Environment, 2017). In 1970, the National Public Health Service acknowledged that African American and Hispanic children were disproportionately impacted by lead poisoning and in 1982, residents of a predominantly African American community in North Carolina protested the construction of a hazardous waste facility in their neighborhood (Maryland Department of the Environment, 2017). In 1987, a national study was published correlating waste facility siting and race. In 1990, the U.S. Environmental Protection Agency founded the Environmental Equity Work Group. In 1991, 17 principles of EJ were set forth at the First National People of Color Environmental Leadership Summit in Washington, D.C (Reprinted by Environmental Working Group, 2007).

17 Principles of Environmental Justice

1. EJ affirms the sacredness of Mother Earth, ecological unity and the interdependence of all species, and the right to be free from ecological destruction.
2. EJ demands that public policy be based on mutual respect and justice for all peoples, free from any form of discrimination or bias.
3. EJ mandates the right to ethical, balanced and responsible uses of land and renewable resources in the interest of a sustainable planet for humans and other living things.
4. EJ calls for universal protection from nuclear testing, extraction, production and disposal of toxic/hazardous wastes and poisons and nuclear testing that threaten the fundamental right to clean air, land, water, and food.
5. EJ affirms the fundamental right to political, economic, cultural and environmental self-determination of all peoples.
6. EJ demands the cessation of the production of all toxins, hazardous wastes, and radioactive materials, and that all past and current producers be held strictly accountable to the people for detoxification and containment at the point of production.
7. EJ demands the right for all peoples to participate as equal partners at every level of decision-making including needs assessment, planning, implementation, enforcement and evaluation.
8. EJ affirms the right of all workers to a safe and healthy work environment, without being forced to choose between an unsafe livelihood and unemployment. It also affirms the right of those who work at home to be free from environmental hazards.
9. EJ protects the right of victims of environmental injustice to receive full compensation and reparations for damages as well as quality health care.
10. EJ considers governmental acts of environmental injustice a violation of international law, the Universal Declaration On Human Rights, and the United Nations Convention on Genocide.
11. EJ recognizes a special legal and natural relationship of Native Peoples to the U.S. government through treaties, agreements, compacts, and covenants affirming sovereignty and self-determination.

12. EJ affirms the need for urban and rural ecological policies to clean up and rebuild our cities and rural areas in balance with nature, honoring the cultural integrity of all our communities, and providing fair access for all to the full range of resources.
13. EJ calls for the strict enforcement of principles of informed consent, and a halt to the testing of experimental reproductive and medical procedures and vaccinations on people of color.
14. EJ opposes the destructive operations of multi-national corporations.
15. EJ opposes military occupation, repression and exploitation of lands, peoples and cultures, and other life forms.
16. EJ calls for education of present and future generations that emphasizes social and environmental issues, based on an appreciation of diverse cultural perspectives.
17. EJ requires that we, as individuals, make personal and consumer choices to consume as little of Mother Earth's resources and to produce as little waste as possible; and make the conscious decision to challenge and reprioritize our lifestyles to insure the health of the natural world for present and future generations.

The EJ movement came of age in 1993-1994, with the passage of the Environmental Justice Act (EJA), the Environmental Equal Rights Act, and the Environmental Health Equity Information Act by Congress, and the establishment of the National Environmental Justice Advisory Council (NEJAC) by the U.S. Environmental Protection Agency. In 1994, President Clinton signed an executive order specifically aimed at addressing environmental injustice (Office of the President of the United States); Executive Order12898 requires that each Federal agency makes the achievement of environmental justice part of its mission. Executive Order12898 was reaffirmed by President Obama in a 2011 Memorandum (Maryland Department of the Environment, 2017).

Today, EJ is an overarching goal of the U.S. Environmental Protection Agency, which defines it as "the fair treatment and meaningful involvement of all people regardless of race, color, national origin, or income, with respect to the development, implementation, and enforcement of environmental laws, regulations, and policies" (U.S. EPA, 2017a). To achieve environmental justice, all people must experience

- The same degree of protection from environmental and health hazards;

- Equal access to the decision-making process to have a healthy environment in which to live, learn, and work (U.S. EPA, 2017a).

ENVIRONMENTAL JUSTICE IN EUROPE

Although the EJ movement began in the United States, it has spread far and wide as the increasingly severe impacts of global warming, combined with the adverse effects from air, water, and soil pollution, have pushed people to fight for a clean, healthy environment, protection from natural hazards, and a voice in the decision- and policy-making process. The Aarhus Convention, signed in 1998 in Aarhus, Denmark, and ratified in 2001, upholds environmental democracy by ensuring citizens' right to be informed about and participate in environmental issues (European Commission, 2016). The European Environmental Bureau (EEB), based in Brussels, Belgium, is the largest federation of environmental citizens' organizations in Europe, with more than 150 member organizations from 30 countries, representing some 15 million individual members and supporters. The EEB describes itself as

the environmental voice of European citizens, standing for environmental justice, sustainable development and participatory democracy. We want the EU to ensure all people a healthy environment and rich biodiversity [EEB, 2017a]

The EEB has initiatives in five work areas:

- **Biodiversity and Nature:** agriculture, biodiversity and nature, soil, and water.
- **Climate and Energy:** biofuel, climate change, and energy policy.
- **Governance and Tools:** Aarhus, enforcement, enlargement, OECD, standardization, and working with presidencies.
- **Industry and Health:** air, chemicals, industrial emissions, mercury, nanotechnology, noise.
- **Sustainability:** 7th environmental action program, ecolabel, ecological product policy, environmental fiscal reform, environmental justice, Europe 2020, Green Public Procurement, sustainable development, and waste [EEB, 2017a].

Biodiversity and Nature

Biodiversity may be defined as the variety and abundance of species on our planet, as well as the genetic diversity within those species, and the diversity of entire habitats and ecosystems (National Wildlife Federation, 2017). Biodiversity includes ecosystems services such as the way wetlands clean water and absorb chemicals; biodiversity also enables ecosystems to adjust to natural hazards such as extreme fires and floods. Genetic biodiversity prevents diseases and helps species to adjust to environmental changes (National Wildlife Federation, 2017).

Under the EU's Biodiversity Strategy 2010-2020, the EU Member States unanimously committed to help stop global biodiversity loss (European Commission, 2017). To work toward achieving this goal, EU Birds and Habitats Directives were enacted; the Birds Directive aims to protect the 500 wild bird species naturally occurring in the European Union, while the Habitat Directive ensures the conservation of a wide range of rare, threatened or endemic animal and plant species (European Commission, 2017). In addition, Natura 2000, the world's largest international network of protected areas, was created. Natura 2000 covers about 20 percent of the EU's land and 6 percent of its seas. The Directives and Natura 2000 address two of the main drivers of biodiversity loss, namely, unsustainable agricultural practices and infrastructure development (European Commission, 2017).

According to the EEB (2017a), Europe's current food and farming system is unsustainable and directly contributes to a wasteful use of finite global resources and damages the environment by contributing to climate change, biodiversity loss, depletion of fisheries, deforestation, soil erosion, water scarcity, as well as water and air pollution. In addition, this system has resulted in widespread undernourishment as well as a rapid increase in obesity, a major cause of disease and death in Europe and worldwide (EEB, 2017a). In a Civil Society Statement on the Reform of European Agricultural Policies (EEB, 2017b), the EEB called for a more sustainable, resilient farming system based on four principles:

- Fair and diverse food and farming economies (decent income and work conditions, short supply chains and sustainable public procurement policies).
- Healthy environment and a food and farming system that respects animal welfare

- Support for citizens' health and well-being (healthy, nutritious, seasonal, local, culturally appropriate and affordable diets with lower consumption of animal products).
- A publicly accountable food system with participatory governance, citizens' empowerment and democracy.

Climate and Energy

The EEB works closely with Climate Action Network Europe in efforts to ensure that the EU fully complies with the Kyoto Protocol (an international agreement linked to the UN Framework Convention on Climate Change, it set internationally binding emission reduction targets beginning with the period 2008-2012) and works to achieve a fair and binding international agreement for the post-2012 period (United Nations, 2014). The EEB is working on environmental tax reform to help fight climate change. The EEB strongly believes that EU climate policy should contribute to the global objective of reducing the rise in global temperatures to well below 2 degrees Celsius (EEB, 2017a).

Governance and Tools

The Aarhus Convention, which entered into force on October 30, 2001, established the public's right to gain access to environmental information held by public authorities, to participate in environmental decision-making, and to access justice to review procedures to challenge public decisions that have been made without respecting the two aforementioned rights or environmental law (European Commission, 2016).

The EEB serves as a liasion between the Organisation for Economic Cooperation and Development (OECD) and environmental non-governmental agencies (NGOs) in OECD countries. The OECD includes EU countries as well as Australia, the United States, Canada, Chile, Iceland, Israel, Japan, Korea, Mexico, and New Zealand. In 2001, OECD Environment Ministers agreed to an environmental strategy that includes these key objectives:

- Maintaining the integrity of ecosystems through the efficient management of natural resources.
- Enhancing the quality of life.
- Global environmental interdependence.

- Improving governance and co-operation (EEB, 2017a).

Industry and Health

Air pollution reduction measures can help to reduce greenhouse gas emissions; likewise, fighting climate change will help to improve air quality. The EEB is part of the "Soot Free for the Climate Campaign" which is a group of European NGOs working together to ensure significant reductions of soot / black carbon emissions in Europe (EEB, 2017a).

The EU must comply with RoHS, which stands for Restriction of Hazardous Substances. RoHS, also known as Directive 2002/95/EC, restricts the use of six hazardous materials found in electrical and electronic products. All applicable products in the EU market after July 1, 2006 must pass RoHS compliance. RoHS impacts the entire electronics industry and many electrical products as well (RoHS, 2017).

Sustainability

The environmental justice movement has created an Environmentalism of the Poor (EOP), which could become a driving force for achieving an ecologically sustainable society. In the past, environmentalism was seen as the preserve of the rich or northern societies while EOP was often overlooked as a movement. The key to empowering EOP is allowing the poor a more equitable and participatory role in sustainable global development.

Europe 2020 lays out the EU's agenda for sustainable growth between now and 2020. It seeks to increase the employment rate, invest more money in research and development, reduce poverty, improve educational completion rates, cut greenhouse gas emissions by 20 percent compared to 1990, increase the amount of energy from renewables to 20 percent of the total, and improve energy efficiency 20 percent over 1990 (EEB, 2017a).

EDUCATIONAL COMPLETION RATES, OECD COUNTRIES AND PARTNERS

Giving all people access to a quality education is a fundamental part of environmental justice that addresses inequalities in educational opportunities in order to improve social mobility and socio-economic outcomes (OECD,

2016). Educational attainment is the completion of a certain level of education, and the holding of a qualification (such as a diploma or certificate) for that level of education. Higher educational achievement is associated with better health, more social engagement, and higher employment rates and relative earnings. Upper secondary school is comparable to senior high school in the U.S.; post-secondary non-tertiary school is vocational school. In Europe, students are tracked for upper secondary school or vocational school by age 15-16 (OECD, 2016). Table 1 shows upper-secondary school first-time graduation rates in the OECD countries and partner countries (far-right column). Also shown are graduation rates for all adults for general programs and vocational programs.

Table 1. Upper secondary (high school) graduation rates for OECD9 countries & partners, 2014

Country	Graduation Rates, General Programs, Adults Aged 25-64	Graduation Rates, Vocational Programs, Adults Aged 25-64	First-Time Graduation Rates, all Programs
Australia	74	80	m
Austria	20	79	90
Belgium	38	55	m
Canada[1]	85	4	89
Chile	59	29	88
Czech Republic	22	57	74
Denmark	68	46	94
Estonia	60	24	m
Finland	46	96	97
France	54	76	m
Germany	48	43	91
Greece	70	33	m
Hungary	66	23	88
Iceland[1]	74	50	89
Ireland	111	45	m
Israel	53	37	90
Italy	38	55	93
Japan	74	23	97
Korea	78	17	95
Latvia	67	27	88

continued on following page

Table 1. Continued

Country	Graduation Rates, General Programs, Adults Aged 25-64	Graduation Rates, Vocational Programs, Adults Aged 25-64	First-Time Graduation Rates, all Programs
Luxembourg	33	43	74
Mexico	33	19	51
Netherlands	42	77	95
New Zealand	76	59	95
Norway	62	37	84
Poland	49	35	83
Portugal	41	56	97
Slovak Republic	27	57	83
Slovenia	36	65	90
Spain	53	29	74
Sweden	48	29	69
Switzerland	42	70	m
Turkey	34	34	68
United Kingdom	m	m	m
United States	m	m	82
OECD Average[2]	50	49	85
EU22 Average[2]	47	55	88
OECD Partners			
Argentina	m	m	59
Brazil	62	6	64
China	47	39	86
Columbia	m	m	70
Costa Rica	m	m	54
India[1]	m	2	m
Indonesia	40	29	69
Lithuania	77	15	92
Russian Federation	52	31	50
Saudi Arabia	m	m	72
South Africa	m	m	34

Notes: m = no data available. 1. Year of reference 2013 instead of 2014. 2. The averages are calculated only from countries with data available for all reference years. EU22 is 21 countries that are both EU and OECD members: Austria, Belgium, the Czech Republic, Denmark, Estonia, Finland, France, Germany, Greece, Hungary, Ireland, Italy, Latvia, Luxembourg, the Netherlands, Poland, Portugal, the Slovak Republic, Slovenia, Spain, Sweden and the United Kingdom.

Source: Table A2.4 (OECD, 2016, p. 59).

The highest-ranking countries for first-time high school graduation rates were Finland, Japan, and Portugal, which were tied for first place with a 97 percent graduation rate. Korea, the Netherlands, and New Zealand were tied for second place with a first-time high school graduation rate of 95 percent. Denmark was next, at 94 percent, followed by Italy at 93 percent, Lithuania at 92 percent, and Germany at 91 percent. Austria, Israel, and Slovenia all had first-time high school graduation rates of 90 percent, bringing the total number of countries with rates of 90 percent or above to 13. The United States was 23rd in the OECD ranking for first-time high school graduation rates, below both the EU22 average of 88 and the OECD average of 85. For all adults 25-64, the graduation rates are considerably lower, especially in Austria (20 percent), the Czech Republic (22 percent), Luxembourg (33 percent), Mexico (33 percent), and Turkey (34 percent) (OECD, 2016).

LITERACY IN DEVELOPING COUNTRIES

Literacy—the ability to read and write-- is a basic requirement for a life above subsistence level, and achieving universal literacy is an important part of cultivating environmental justice. In the developed nations of the world, literacy is now substantially universal. However, in sub-Saharan Africa, the most recent available data showed a literacy rate of only 59 percent in 2011, with more than one-third of adults unable to read and write. In South and West Asia, the adult literacy rate was only 63 percent, and in the Caribbean, it was only 69 percent for the same period (Huebler & Lu, 2013).

Three-quarters of the global illiterate population lived in South and West Asia and sub-Saharan Africa. South and West Asia had 407 million illiterate adults, more than 50 percent of the global total, and sub-Saharan Africa had 182 million illiterate adults, almost 25 percent of the worldwide total. In other regions for which there were data, the adult illiterate population was as follows: East Asia and the Pacific (89 million), Arab States (48 million), Latin America and the Caribbean (36 million), Central and Eastern Europe (5 million), and Central Asia (0.3 million) (Huebler & Lu, 2013).

Data showed a gender gap between men and women. Except in Central Asia, women had higher rates of illiteracy than men. The gender gap was especially large in the Arab States; the male literacy rate was 85 percent versus a female rate of 68 percent, Worldwide, a staggering 493 million women were illiterate; that is, two-thirds of the global illiterate population were women (Huebler & Lu, 2013). Environmental justice demands equal access to education, without fear of violent acts such as the April 2014 kidnapping of some 270 schoolgirls in Chibok, Nigeria by the Islamist militant group Boko Haram.

ADDRESSING HEALTH DISPARITIES TO CULTIVATE ENVIRONMENTAL JUSTICE

Working toward environmental justice means addressing inequities in multiple arenas of daily life, including socioeconomic status, education, safety, food security, access to clean water and sanitation, and access to affordable healthcare. When one discusses the health impacts of climate change, it is imperative to address the underlying health disparities. The Centers for Disease Control and Prevention defines health disparities as "preventable differences in the burden of disease, injury, violence, or opportunities to achieve optimal health that are experienced by socially disadvantaged populations" (CDC, 2015). Some of the factors that create health disparities include:

- Poverty;
- Environmental threats;
- Inadequate access to health care;
- Educational inequalities (CDC, 2015).

HEALTHY PEOPLE 2020 AND ENVIRONMENTAL JUSTICE

Healthy People provides a blueprint for health for the United States; created by a bank of health experts, it provides science-based, 10-year national objectives for improving the health of all Americans. *Healthy People* has established benchmarks and monitored progress, encouraging collaborations across communities and sectors, empowering individuals to make informed health decisions, and evaluating the impact of prevention activities. *Healthy People 2000* focused on reducing health disparities, while *Healthy People 2010* sought to eliminate, not just reduce, health disparities. The overarching goals of *Healthy People 2020* are as follows:

- *Attain high-quality, longer lives free of preventable disease, disability, injury, and premature death.*

- *Achieve health equity, eliminate disparities, and improve the health of all groups.*
- *Create social and physical environments that promote good health for all.*
- *Promote quality of life, healthy development, and healthy behaviors across all life stages (Healthy People 2020, 2017a).*

Healthy People 2020 has set targets to meet the goals set forth in the U.S. Department of Health and Human Services' (HHS) Environmental Justice Strategy and Implementation Plan (U.S. HHS, 2012). When the HHS Plan was issued in 2012, then Secretary Kathleen Sebelius opened it with a message:

With the release of our Strategy and Implementation Plan, we are renewing our commitment to working with our Federal partners to promote environmental justice. (Sebelius, 2012)

Sebelius emphasized the extreme importance of a healthy environment to the national economy:

Every American deserves to have a clean, safe and healthy environment. Today, we understand better than ever before that our health is not only dependent on what happens in the doctor's office but is determined by the air we breathe, the water we drink and the communities we call home….There is nothing more important than health. Health is the foundation of our nation's prosperity. (Sebelius, 2012)

Sebelius went on to note that the CDC had awarded $103 million in community transformation grants to state and local government agencies, Tribes, territories, and nonprofits to promote healthy lifestyles, especially among vulnerable population groups experiencing the greatest burden of chronic disease. A major goal of these grants was to reduce health disparities (U.S. HHS, 2012).

Healthy People 2020's environmental justice initiative has a total of 310 objectives, divided into four topics: 1) access to health services, 2) educational and community-based programs, 3) environmental health, and 4) public health infrastructures.

Access to Health Services

The first topic area has 32 objectives. The first objective (AHS-1) focuses on increasing the numbers of people with health insurance, dental insurance, and prescription drug insurance. This last category is important for chronic disease management. For example, historically some Americans with asthma had health insurance that capped their prescription drugs at a certain level; once that level was exceeded, medications were not covered. Asthmatics would run out of coverage for their medications. Adequate coverage of prescription drugs is essential for health equity.

The first topic area also focuses on increasing health insurance coverage for clinical preventive services (AHS-2), increasing the number of people who have an ongoing primary care provider (AHS-3), as well as the number of practicing primary care providers (AHS-4), and sources of primary care by age (AHS-5). Other objectives include how long it takes to receive medical and dental and prescription drug care (AHS-6), receipt of evidence-based clinical preventive services (AHS-7), rapid prehospital emergency care (AHS-8), and wait time at emergency departments (AHS-9) (*Healthy People 2020*, 2017b). Without adequate clinical preventive services and environmental controls in the home and neighborhood, poor people wind up with severe asthma exacerbations and breathing emergencies. America's EDs are overburdened by a sizable number of patients who could be seen elsewhere (Sommers et al., 2012). The National Health Statistics Report (2015) noted that approximately 20 percent of U.S. adults seek treatment in EDs every year, a percentage that has remained consistent over the past ten years. The heaviest users of EDs are those with Medicaid, who tend to be in poorer health overall than other populations (Tang et al., 2010; National Center for Health Statistics, 2013; Sommers et al., 2012; Adams, 2013).

The emergency medical system is stretched beyond capacity. It would be beneficial to develop less-costly care options than emergency departments, such as urgent care settings that can quickly handle a high volume of medical problems (Sommers et al., 2012). One area that potentially could be served outside EDs is some acute respiratory and other infections in children and injuries; together these diagnoses account for about 53 percent of ED visits by children aged 0 to 12 covered by Medicaid and nearly 60 percent of ED visits by privately insured children aged 0 to 12 (Sommers et al., 2012). The overburdened state of EDs could be deadly when a large-scale disaster strikes. In most states, the hospital system would be overwhelmed trying to absorb the surge in demand associated with a pandemic, natural disaster, or terrorist attack (Shea, 2007).

Educational and Community-Based Programs

The second topic area of environmental justice, educational and community-based programs, has 122 objectives. *Healthy People 2020* has included ECBP-6, high school completion, as one of the leading health indicators for Americans because it predicts future health status and employment status. Education is one of the strongest predictors of health: the more schooling people have the better their health is likely to be (Freudenberg & Ruglis, 2007). In addition, the less schooling people have, the higher their levels of risky health behaviors such as smoking, being overweight, or having a low level of physical activity (Lantz et al., 1998). Figure 1 shows the 2013-2014 adjusted cohort graduation rate (ACGR) for public high school students in the United States.

Figure 1. Adjusted cohort graduation rate (ACGR) for public high school students, by state: School year 2013–14

Note: The adjusted cohort graduation rate (ACGR) is the percentage of public high school freshmen who graduate with a regular diploma within 4 years of starting 9th grade. The Bureau of Indian Education and Puerto Rico are not included in the U.S. 4-year ACGR estimate. All rates are rounded to nearest whole number.

Source: U.S. Department of Education, Office of Elementary and Secondary Education, Consolidated State Performance Report, 2013–14. See Digest of Education Statistics 2015, Table 219.46.

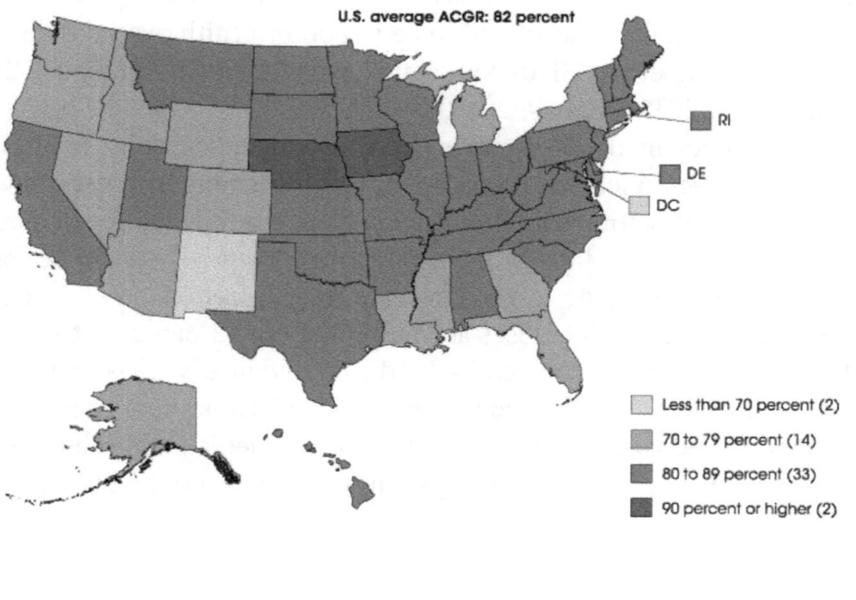

The top indicators for high school completion within four years after 9th grade are attendance, GPA, and test scores; in particular, missing no more than 10 percent of school days per grade level is associated with on-time high school graduation (Allensworth & Easton, 2007). Asthma affects one in 11 children and is a leading cause of school absenteeism in the United States, causing 10.5 million missed school days annually (Centers for Disease Control and Prevention, 2013; Krenitsky-Korn, 2011). Poorly controlled asthma leads to acute exacerbations, emergency department visits, and hospitalizations (Evans et al., 1997; Lara et al., 2002). However, with adequate asthma management by a clinician, children with asthma can attend school regularly and graduate. The key person responsible for school health is the school nurse, and most schools do not have the recommended full-time registered school nurse-to-student ratio of 1:750. In 2006, only 33.5 percent of high schools, 43.9 percent of middle schools, and 41.4 percent of elementary schools met this target. Objective ECBP-5 seeks to increase by 10 percent the proportion of elementary, middle, and senior high schools that have a full-time registered school nurse-to-student ratio of at least 1:750 by 2020 (*Healthy People 2020*, 2017b).

Other objectives addressed by the educational and community-based programs are health education programs that teach prevention to preschool (ECBP-1) school-age (K-12) (ECBP-2) and college students (ECBP-7). These health education objectives cover prevention of unintentional injury, violence, tobacco use and addiction, alcohol or other drug use, unhealthy dietary patterns, inadequate physical activity, dental health, and safety (*Healthy People 2020*, 2017b). Objective ECBP-3 addresses the implementation of National Health Education Standards for elementary, middle school, and high school students, in the areas of knowledge (understanding health issues), skills (being able to access health information and services), health advocacy (for themselves, their family and community), evaluating health information in the media, using interpersonal skills to enhance health, practicing goal setting and decision-making regarding their health behaviors, and putting into practice healthy behaviors (*Healthy People 2020*, 2017b). Objective ECBP-4 focuses on health education in school-age children to improve sanitation--hand washing or hand hygiene--oral health, growth and development, sun safety and skin cancer prevention, sleep habits, and adherence to checkups and health screenings; it also deals with preventing vision and hearing loss (*Healthy People 2020*, 2017b).

The remainder of the objectives for education deal with including core clinical prevention and population health content in the training of healthcare providers such as MDs, RNs, DOs, and others, and increasing the proportion of academic institutions with health professions education programs whose prevention curricula include interprofessional educational experience (*Healthy People 2020*, 2017b). Two objectives deal with increasing both the number of worksites that offer health promotion programs and the number of employees who participate in these programs (ECBP-8 and ECBP-9). Community-based objectives (ECBP-10 ECBP-11) seek to increase the number of local health departments, Tribal health services, nongovernmental organizations, State agencies, and community organizations that provide population-based primary prevention services in the areas of injury and violence, mental illness, tobacco use, substance use, unintended pregnancy, chronic disease, nutrition, and physical activity. Also targeted is an increase in the proportion of health departments that have established culturally appropriate and linguistically competent community health promotion and disease prevention programs (*Healthy People 2020*, 2017b)

Environmental Health

The third topic area, environmental health, has 77 objectives. The first three deal with outdoor air quality. EH-1 focuses on the air quality index (AQI) not exceeding 100, while EH-2 deals with using alternative modes of transportation rather than traditional internal combustion engine vehicles, which are a major source of outdoor air pollution. Alternate modes of transportation targeted by this initiative are bicycling, walking, using mass transit, and telecommuting. Objective EH-3 focuses on reducing toxic airborne emissions. On December 15, 2016, the U.S. Department of Energy (DOE) announced a $19.7 million initiative to support research and development of advanced vehicle technologies, including batteries, lightweight materials, and advanced combustion engines, as well as innovative technologies for energy efficient mobility (U.S. DOE, 2017). Progress in the United States has been slow, as each state has to make its own decision about EVs.

Initiatives to switch over from internal combustion engine vehicles to electric vehicles have gathered momentum in Europe. The International Alliance for Zero-Emission Vehicles (ZEV) was founded by 11 partners in 2011 to transition to clean electric vehicles with the aim of limiting the impact of climate change (Barnum, 2015). The founding partners were The Netherlands, Norway and the United Kingdom in Europe; California, Connecticut, Maryland, Massachusetts, Oregon, Rhode Island and Vermont in the United States; and Québec (Barnum, 2015). In 2016, Germany joined the International Alliance for ZEV and launched an incentive (4,000 Euro discount for EVs) and investment program to accelerate the development of EVs and switch all vehicle production to EVs by 2030 (Lambert, 2016). However, the market share for EVs is still very tiny there. In 2015, Norway was far and away the world leader in EV vehicles, with EV penetration of 23.3 percent (IEA, 2016). The Netherlands was a distant second, with an EV penetration of 9.7 percent (IEA, 2016). Sweden was third, with an EV market penetration of 2.4 percent, followed by France at 1.2 percent, China and the UK at 1.0 percent, and Germany, Portugal, and the United States at .07 percent (IEA, 2016). Germany has adopted an evidence-based theoretical framework (Diffusion of Innovations—see Chapter 5) and incentivized marketing campaign to promote the adoption of EV vehicles, so its EV penetration is likely to increase substantially in the near future.

Objectives EH-4 and EH-5 of environmental health focus on water safety. The former focuses on increasing the proportion of people drinking water from community water systems that meets the standards of the Safe Drinking Water Act, and the latter seeks to reduce outbreaks of water-borne infections from 7 per year to 2 per year. Water conservation, via usage reduction, is the focus of objective EH-6, and keeping beaches open and safe more days per year is the focus of objective EH-7.

The next five objectives deal with toxics and waste, from reducing lead blood levels in children (EH-8) to minimizing risks to humans and the environment posed by hazardous sites (EH-9), reducing pesticide exposures (EH-10) and toxic pollutants released into the environment (EH-11), and increasing recycling of municipal solid waste (EH-12).

The next six objectives of environmental health focus on healthy homes and communities, from reducing indoor air allergens in homes from cockroach and mouse droppings (E-13), to reducing and mitigating radon in homes (EH-14 and EH-15), increasing testing of pre-1978 housing for lead-based paint or related hazards and reduce number of homes with lead problems (EH-17 and EH-18), and reducing the proportion of occupied housing units that have moderate or severe physical problems (EH-19). Objective EH-16 focuses on increasing the proportion of elementary, middle and high schools that have official school policies and engage in practices that promote a healthy and safe physical school environment, including indoor air quality management programs, mold remediation, safe drinking water, safe disposal of hazardous waste, and pest management systems that minimize human exposure (*Healthy People 2020*, 2017b).

The following objectives of environmental health are for infrastructure and surveillance: reducing exposures to selected chemicals in the population, as measured by blood and urine of the substances or their metabolites (EH-20), improving information systems that monitor environmental health (EH-21), increasing the number of States, Territories, Tribes, and the District of Columbia that monitor diseases or conditions that can be caused by exposure to environmental hazards (EH-22), and reducing the number of public schools located within 150 meters (164 yards) of major highways in the United States (EH-23). There is one last goal for the environmental health topic area, namely, reducing the global burden of disease due to poor water quality, sanitation, and insufficient hygiene (EH-24) (*Healthy People 2020,* 2017b)

Public Health Infrastructures

The fourth topic area, public health infrastructures, has 79 objectives. The first objective (PH-1) focuses on increasing the proportion of Federal, Tribal, State, and local public health agencies that incorporate Core Competencies for Public Health Professionals into job descriptions and performance evaluations, and the following three objectives focus on including Core Competencies in Continuing Education for Public Health Professionals, and increasing the number of colleges and universities that offer majors and minors in Public Health. Objectives PH-7 through PH-10 deal with improving data collection and information systems in order to gather more thorough and complete data on the population-based *Healthy People 2020* objectives. The next two objectives (PH-11 and PH-12) focus on increasing the proportion of Tribal and State public health agencies that provide comprehensive laboratory services that perform at a high level of quality in support of the 10 Essential Public Health Services, which are:

- Monitoring health status to identify and solve community health problems;
- Diagnosing and investigating health problems and health hazards in the community;
- Informing, educating, and empowering people about health issues;
- Mobilizing community partnerships and action to identify and solve health problems;
- Developing policies and plans that support individual and community health efforts;
- Enforcing laws and regulations that protect health and ensure safety;

- Linking people to needed personal health services;
- Assuring a competent public and personal health care workforce;
- Evaluating effectiveness, accessibility, and quality of personal and population-based health services;
- Supporting research into new insights and innovative solutions to health problems.

The remainder of the objectives (PH-13 through PH-17) focus on increasing the proportion of Tribal, State, and local public health agencies that have a health improvement plan linked to their state plan, that provide comprehensive epidemiology services, that perform public health assessments in accordance with national standards, that have implemented an agency-wide quality improvement process, and that are accredited.

U.S. ENVIRONMENTAL PROTECTION AGENCY EJSCREEN TOOL

In keeping with its goal of cultivating environmental justice by empowering communities to work toward achieving environmental justice, the U.S. Environmental Protection Agency has developed a new environmental justice (EJ) mapping and screening tool called EJSCREEN; it is based on national data and combines environmental and demographic indicators in maps and reports (U.S. EPA, 2017b). This new technology enables any user with an Internet connection to enter a specific location (block, tract, or region) and then look up demographic indicators such as percentages of different racial and ethnic groups, household income, mean price of home, percentages of homeowners versus renters, language(s) spoken in home, educational level, and gender and ages of residents. By clicking on "Printable Standard Report," EJScreen users can access a table of data showing pollution levels of particulate matter and ozone, as well as the National Air Toxins Assessment (NATA) Air Toxics Cancer Risk and Respiratory Hazards Index, Traffic Proximity and Volume, Lead Paint Indicator, and proximity to Superfund, RMP, Hazardous Waste, and Water Discharger (U.S. EPA, 2017b). Pollution levels are shown for the regional, state, and national levels for comparative purposes (U.S. EPA, 2017b).

GEORGRAPHIC INFORMATION SYSTEM (GIS) MAPPING TOOL

Another innovative technology that is useful for visually depicting data on environmental justice problems is Geographic Information System (GIS) software. GIS Software is designed to store, retrieve, manage, display, and analyze all types of geographic and spatial data. Users of the GIS software can produce maps and other graphic displays of geographic information for analysis and presentation. For example, users who are surveying a community for water issues may use GIS Software to map the percentages of homes in a given area that have experienced flooding, mold, and water contamination (ReactToolkit.net, 2017). The U.S. Environmental Protection Agency distributes small grants that fund pilot studies on environmental justice. One such study, in Palm Beach County, Florida, surveyed two vulnerable coastal communities for water issues and then used GIS Software to visually map findings on reports of flooding (blue teardrop), blocked storm drains (orange x), and flooded cars (red car). One of these GIS maps for a primarily African-American neighborhood in southeast Boca Raton, Florida is shown below (ReACTtoolkit.net, 2017).

GIS uses geocoding to find locations worldwide; then users input data on indicators, which are depicted in layers on the map. An array of formats and colors are available so that multiple indicators can be shown on one map. Geocoding and inputting data on geomaps may be used to prioritize need and justify investment in specific community redevelopment projects. Geomapping requires some software expertise as well as an outlay of funds to purchase the software. A free one-month training program in geomapping, Do-It-Yourself GeoApps, is available on line as a Massive Online Open Course (MOOC) (Esri.com, 2017). It enables potential users to try out GIS software for free before purchasing it.

CASE EXAMPLES

The following case studies in environmental justice are examples of how a community can advocate for and address longstanding problems with EJ issues such as severe water contamination caused by pesticides and extreme air pollution caused by improper industrial waste management.

Figure 2. GIS Map of Southeast Boca Raton
Source: Rising Together, U.S. EPA Environmental Justice Project (ReACTtoolkit.net, 2017)

Case Example 1: DDT Contamination in Triana, Alabama

In 1979, the Tennessee Valley Authority (TVA) reported extensive DDT contamination in the water of the Huntsville Spring Branch near the small town of Triana, Alabama. To understand the importance of this case, a brief history of DDT is in order.

DDT (dichloro-diphenyl-trichloroethane) was developed in the 1940s as a synthetic insecticide and was used to combat malaria and other insect-borne diseases among military and civilian populations, as well as for insect control in crop and livestock production, institutions, homes, and gardens (U.S. EPA, 2016). Marine biologist Rachel Carson's 1962 book *Silent Spring* alerted the public to the dangers of how DDT disrupted natural systems, and was stored in the fatty tissues of animals (including human beings), causing cancer and genetic damage (Carson, 1962). There was much opposition to Carson's book, as DDT was a very profitable product. Carson testified before Congress in 1963 and was fiercely opposed by the chemical industry. Sadly, she did not live to see her work validated; she died of cancer in 1964 at age 57. In 1972, EPA banned the use of DDT, based on its adverse environmental effects and suspected carcinogenic properties (U.S. EPA, 2016).

Triana, Alabama is a largely poor, predominantly African-American community whose residents rely heavily on fish from the Spring Branch for food. The TVA tested fish caught in the Spring Branch and found DDT levels as high as 200 parts per million (ppm), 40 times the federal limit (Rebitzke, 2004). The presence of DDT in the water was linked to the Redstone Arsenal, located in nearby Huntsville, Alabama. The Army Corps of Engineers owned the Redstone Arsenal and leased it to the Olin Corporation from 1947 until it was closed in 1970 (Rebitzke, 2004).

After the TVA report was released, the mayor of Triana, Clyde Foster, requested that residents be checked for signs of DDT in their bodies. The Centers for Disease Control and Prevention (CDC) initially tested twelve Triana residents for DDT in their blood, and found levels three times the normal levels of DDT found in case studies of workers at DDT plants. None of the twelve Triana residents who were tested by the CDC had ever worked at such a facility. The CDC went on to test 500 residents of Triana (Rebitzke, 2004). The EPA began an investigation in order to determine how the DDT accumulated in the area, and it found that the source of the DDT was the Redstone Arsenal, which manufactured DDT from 1947-1970. After the results of the investigation were released, the Army Corps of Engineers was ordered to clean up the DDT in the surrounding area and to test the local residents for possible contamination. However, the Army Corps refused to do so, stating that they were not responsible for any contamination that was present outside the facility boundaries. In 1980, all the residents of Triana, along with the Justice Department, filed a lawsuit against the Olin Corporation (Rebitzke, 2004).

In 1982, the Olin Corporation made an initial offer of a flat fee per plaintiff found to have elevated DDT blood levels. It was not accepted and further lawsuits ensued. After four more years, in May 1986, approximately one month before trial, all parties agreed to a $15,000,000 settlement to be paid over two years. In addition, Olin would be required to clean up the DDT in the area at their own expense, which could be as high a $137 million (Rebitzke, 2004). To oversee the DDT cleanup of the Huntsville Spring Branch, an Interagency Review Panel was established; it comprised the U.S. Environmental Protection Agency, the TVA, the U.S. Fish and Wildlife Service, the Department of the Army, and the State of Alabama. For ten years the Interagency Review Panel worked with Olin to reduce DDT levels in the fish in the Harbor Branch to normal levels and to clean up the areas surrounding the Redstone Arsenal. Between 1985 and 1995, the Panel monitored the Triana cleanup efforts; the DDT levels in the major fish species in Harbor Branch returned to normal and the DDT levels in the waters surrounding the Redstone Arsenal and the town of Triana were reduced by 97 percent (Rebitzke, 2004).

After the total settlement amount was agreed upon in May 1986, blood tests and other information on plaintiffs were obtained by the court (McGovern, 1990). The judge ruled on the amount per plaintiff in November 1988, more than nine years after the case was initiated. The settlement amount per plaintiff was not as high as they had hoped for, in part because the settlement was used to pay extensive legal fees and court administrative costs (McGovern, 1990). However, it was a big win for environmental justice and showed extreme persistence in the face of opposition from a major corporation and the U.S. Army Corp.

Case Example 2: Toxic Doughnut in Chicago

Altgeld Gardens, a 99 percent minority public housing development located in the Calumet industrial area in southeast Chicago, was established in 1945 as a federal housing project for World War II African American veterans (Hintzen, 2015). In 2015, according to EJScreen (U.S. EPA, 2017b), its per capita income was $11,515, below the poverty threshold of $12,082 established by the Social Security Administration and updated yearly by the U.S. Census Bureau (Institute for Research on Poverty, 2016). The community has the highest percentage of people living in poverty and the lowest per capita income in the city (City of Chicago, 2015). Altgeld Gardens was built on an abandoned waste site and is surrounded by the most landfills per square mile in the United States (Pellow, 2002). Altgeld gardens was labelled "the toxic doughnut" because it was ringed by 50 landfills and 382 industrial facilities, and had 250 leaking underground storage tanks (Hintzen, 2015). Historically, African American communities have been disproportionately exposed to environmental hazards (Bullard et al., 2007). In 1998 an estimated 26,000 pounds of toxic pollutants were emitted annually, which contributed to the neighborhood's extremely high rates of respiratory illness and asthma (Hintzen, 2015). Proximity to hazardous facilities and exposure to air pollution are both strong contributors to incidence and mortality rates from stroke (Brook et al., 2014). In 2015, Altgeld Gardens had some of the highest mortality rates for lung cancer and cerebrovascular disease (stroke) in the city (City of Chicago, 2015).

The Altgeld Gardens case is one of the first to document and address environmental racism in a community. The driving force behind the Altgeld Gardens case was Hazel M. Johnson, a resident of the public housing who founded a not-for-profit community-based environmental organization, People for Community Recovery (PCR), in 1979 (Chicago Public Library, 2009). Johnson's interest in environmental justice grew out of her own family's experiences with air pollution in the neighborhood; her husband John, a relatively young man who rarely smoked, died of lung cancer in 1969. The Johnsons' seven children, then aged 2 to 16 years, suffered from skin and respiratory ailments (Chicago Public Library, 2009).

Johnson was raised in New Orleans and met her husband there. They moved to Chicago in the mid-1950s and to Altgeld Gardens in 1962. The public housing development was managed by the Chicago Housing Authority (CHA), and was home to approximately 10,000 residents, mostly low-income minority families (Chicago Public Library, 2009). Johnson worked as a mail sorter for the U.S. Postal Service from 1965 to 1967, as a typist, file clerk and receptionist for Continental Temporary Services from 1968-1969, and as a receptionist and general office support at Parents and Friends of Retarded Children in 1969 (Chicago Public Library, 2009). After her husband's death, Johnson became worried about the polluted environment of Altgeld Garden. She noticed that factory chimneys were pouring out noxious fumes while the ground was filled with toxic waste dumps. There was a constant stream of foul emissions from the air, water, and soil of Altgeld Gardens.

After Johnson saw a TV program about environmentally-related cancer, she became an activist and researcher with the goal of reducing environmental pollution in her neighborhood (Chicago Public Library, 2009). She discovered that Altgeld Gardens had been built upon a toxic waste dump and sewage farm, and that her neighbors and friends had far more health problems than the norm. Johnson spoke out about the sad state of the public housing development infrastructure, which the Chicago Housing Authority (CHA) did not maintain properly or repair; nor did the CHA address the fumes from the toxic landfill underneath the development (Chicago Public Library, 2009). In 1970 Johnson ran for office as block captain and earned a seat on the Altgeld Gardens Local Advisory Council (LAC), a position she held until 1979. Johnson identified the major polluters surrounding Altgeld Gardens, such as Waste Management, Sherwin-Williams Paint Co., PMC Specialty, Ford Motor Co., and the Metropolitan Water Reclamation District of Chicago, and notified her neighbors (Chicago Public Library, 2009). In 1979, PCR was founded to fight environmental racism and reduce exposure to pollutants for the residents of

Altgeld Gardens public housing development. PCR educated residents about tenants' rights and the health hazards of pollution, and supported actions against polluters and negligent government bureaucrats (Chicago Public Library, 2009). In the early days of PCR, Johnson was often the only person of color at environmental conferences. By 1992, the movement had become more diverse and inclusive, thanks in part to her leadership. When she attended the first National People of Color Environmental Leadership Summit in 1992, Johnson was dubbed "Mother of the Environmental Justice Movement." By then, PCR had an executive director (Johnson), an administrative assistant, four outreach workers, thirteen board members, 200 dues-paying members, volunteers, and a $125,890 budget funded largely by grants (Chicago Public Library, 2009).

PCR conducted surveys with Altgeld Garden residents on their health status, starting with a cancer survey in 1980. The survey data showed abnormally high rates of infant death, cancer, respiratory, pulmonary, and skin diseases among Altgeld Gardens residents (Chicago Public Library, 2009). PCR took to the streets with the well-publicized 1987 blockade of the gate of CID Waste Management to prevent some 57 dump trucks from discharging into the landfill. Other social protests led to permit revocations for the Paxton Lagoon and the O'Brien Lock and Dam Landfills. Lobbying the U.S. Environmental Protection Agency, testifying at hearings, and filling out hundreds of anti-landfill applications finally resulted in the 1992 consent decree in which Waste Management agreed to hire Altgeld residents to monitor compliance. This monumental step ended the opening of new landfills and incinerators in the neighborhood (Chicago Public Library, 2009). Johnson was asked to sit on the newly formed National Environmental Justice Advisory Council (NEJAC), which provided a federal-level forum for grassroots representatives to be heard. She travelled abroad to promote environmental justice in Brazil as part of the United Nations Earth Summit in 1992 and to South Africa as part of a 12-member People of Color delegation in 1996 (Bullard, 2011). In 1998-1999, PCR persuaded the U.S. Environmental Protection Agency to require the CHA to remove PCB-contaminated soil from Altgeld Gardens.

Johnson skillfully used the media to influence public opinion; PCR published articles in the *Chicago Sun-Times*, invited the public TV station to film a Town Meeting, and started its own newsletter, *Fighting Against a Toxic Environment (F.A.T.E.)*, with articles written by professional environmentalists and EJ leaders. In 2000-2001, two documentaries on PCR's work on environmental pollution were created and broadcast by National Public Radio and by Northwestern University. In 2000, the management company of Altgeld Gardens was cited by OSHA, ILEPA and the City of Chicago Department of the Environment for improper removal of asbestos contained floor tiles from a Day Care Center (Chicago Public Library, 2009).

Hazel Johnson died in January 2011. Her daughter Cheryl Johnson took over the helm of PCR. In 2013, PCR forced the Chicago Housing Authority (CHA) to contract with an urban planner to develop a master plan for the finishing of the Altgeld redevelopment plan.

Dr. Robert Bullard, Professor of Urban Planning and Environmental Policy at Southern Texas University, and an expert on environmental racism, eulogized Johnson as follows:

Today, Altgeld Gardens is cleaner and healthier for her work. And because of her efforts, some of the dirtiest nearby industrial sites have cleaned up their acts. Much work is still needed.... The EJ Movement and the world will miss this great warrior and healer. Her legacy lives on in her work and the thousands of lives she has touched the nearly 76 years she was with us (Bullard, 2011).

The Altgeld Gardens case is described in the EJ Atlas (Hintzen, 2015) as one of the most influential EJ cases in the United States. Hazel Johnson is widely regarded as one of the founders of the environmental justice movement and an inspiring leader in the fight against environmental racism.

CONCLUSION

Our case examples showed how even the poorest communities can empower their residents to fight for and achieve some measure of environmental justice. However, the roots of environmental injustice go deep, intertwined with longstanding inequities in the treatment of poor communities of color, which historically have been subject to high levels of unsafe dumping of toxic wastes and whose children are exposed to both indoor and outdoor air pollution and toxins, with resulting health problems. The problems of poverty and illiteracy must be remedied; access to adequate sanitation, clean water, and food security are all vital to the achievement of EJ.

A healthy environment in which to live, work, study, and play is a fundamental right of all people. In addition, it has become clear both domestically and abroad, that sustainable development with clean, renewable energy sources is the bedrock of a healthy future for all the world's citizens. Sustainable development is addressed in more detail in Chapter 6.

REFERENCES

Adams, J. G. (2013). Emergency department overuse: Perceptions and solutions. *Journal of the American Medical Association, 309*(11), 1173–1174. doi:10.1001/jama.2013.2476 PMID:23512065

Allensworth, E. M., & Easton, J. Q. (2007). What matters for staying on track and graduating in Chicago Public High Schools: A close look at course grades, failures, and attendance in the freshmen year. Chicago: Consortium on Chicago School Research. Retrieved from http://ccsr.uchicago.edu/sites/default/files/publications/07%20What%20Matters%20Final .pdf

Barnum, A. (2015). International Alliance on Zero-Emission Vehicles Grows to 11 Partners Air Resources Board: California EPA. Retrieved from https://www.arb.ca.gov/newsrel/newsrelease.php?id=761

Brook, R. D., Franklin, B., Cascio, W., Hong, Y., & Howard, G. (2014). Air pollution and cardiovascular disease: A statement for healthcare professionals from the expert panel on population and prevention science of the American Heart Association. *Circulation, 109*(21), 2655–2671. doi:10.1161/01.CIR.0000128587.30041.C8 PMID:15173049

Bullard, R. (2011). Environmental Justice Movement Loses Southside Chicago Icon Hazel Johnson. *OpEd News*. Retrieved from https://www.opednews.com/articles/Environmental-Justice-Move-by-Robert-Bullard-110114-323.html

Bullard, R., Mohai, P., Saha, R., & Wright, B. (2007). *Toxic wastes and race at twenty: 1987– 2007. Grassroots struggles to dismantle environmental racism in the United States*. Cleveland, OH: United Church of Christ Justice and Witness Ministries.

Carson, R. (1962). *The Silent Spring*. Boston, New York: Houghton Mifflin.

Centers for Disease Control and Prevention (CDC). (2013). Asthma's impact on the nation. Data from the CDC National Asthma Control Program. Retrieved from www.cdc.gov/asthma/impacts_nation/default.htm

Centers for Disease Control and Prevention (CDC). (2015). Health disparities. Atlanta, GA: U.S. Department of Health and Human Services. Retrieved from https://www.cdc.gov/healthyyouth/disparities/

Chicago Public Library. (2009). People for community recovery archives. Biography: Hazel M. Johnson. Retrieved from https://www.chipublib.org/fa-people-for-community-recovery-archives/

City of Chicago. (2015). Data portal. Retrieved from https://data.cityofchicago.org/

Environmental Working Group. (2007). 17 Principles of Environmental Justice. Reprinted from 1991 declaration at National People of Color Leadership Summit, Washington, D.C. Retrieved from http://www.ewg.org/enviroblog/2007/10/17-principles-environmental-justice

Esri.com. (2017). Do-It-Yourself GeoApps. MOOC. Retrieved from http://www.esri.com/mooc/diy-geo-apps

European Commission. (2016). The Arhus Convention. Brussels, Belgium: European Commission. Retrieved from http://ec.europa.eu/environment/aarhus/

European Commission. (2017). Environment. Brussels, Belgium: European Commission. Retrieved from http://ec.europa.eu/environment/nature/legislation/birdsdirective/index_en.htm

European Environmental Bureau (EEB). (2017a). We strive for environmental justice. Retrieved from http://www.eeb.org/

European Environmental Bureau (EEB). (2017b). Civil Society Statement on the Reform of European Agricultural Policies Good Food, Good Farming – Now! Retrieved from http://www.eeb.org/index.cfm/library/civil-society-statement-on-the-reform-of-european-agricultural-policies-good-food-good-farming-now/

Evans, D., Mellins, R., Lobach, K., Ramos-Bonoan, C., Pinkett-Heller, M., Wiesemann, S., & Clark, N. (1997). Improving care for minority children with asthma, professional education in public health clinics. *Pediatrics*, *99*(2), 157–164. doi:10.1542/peds.99.2.157 PMID:9024439

Freudenberg, N., & Ruglis, J. (2007). Reframing school dropout as a public health issue. *Preventing Chronic Disease: Public Health Research, Practice, and Policy, 4*(4). Retrieved from https://www.cdc.gov/pcd/issues/2007/oct/pdf/07_0063.pdf

Healthy People 2020. (2017a). Overarching goals. Washington, D.C.: U.S. Department of Health and Human Resources. Retrieved from https://www.healthypeople.gov/2020/About-Healthy-People

Healthy People 2020. (2017b). Environmental justice. Washington, D.C.: U.S. Department of Health and Human Resources. Retrieved from https://www.healthypeople.gov/2020/prevention-portal-508/initiative/hhs-environmental-justice-strategy

Hintzen, K. (2015). Chicago's Toxic Doughnut. Ann Arbor, MI, U.S.A.: University of Michigan; Retrieved from http://ejatlas.org/conflict/chicagos-toxic-doughnut-usa

Huebler, F., & Lu, W. (2013). UNESCO Institute for Statistics. Adult and youth literacy. National, regional, and global trends, 1985-2015. Retrieved from http://www.uis.unesco.org/Education/Documents/literacy-statistics-trends-1985-2015.pdf

Institute for Research on Poverty. (2016). *What are poverty thresholds and poverty guidelines?* Madison, WI: Institute for Research on Poverty.

International Energy Agency (IEA). (2016). Global EV Outlook 2016: Beyond one million electric cars. Paris, France: OECD/IEA, Electric Vehicles Initiative. Retrieved from http://www.iea.org/publications/freepublications/publication/Global_EV_Outlook_2016.pdf

Krenitsky-Korn, S. (2011). High school students with asthma: Attitudes about school health, absenteeism, and its impact on academic achievement. *Pediatric Nursing*, *37*(2), 61–68. PMID:21661604

Lambert, F. (2016). All new cars mandated to be electric in Germany by 2030. Cupertino, CA: Electrek. Retrieved from https://electrek.co

Lantz, P. M., House, J. S., Lepkowski, J. M., Williams, D. R., Mero, R. P., & Chen, J. (1998). Socioeconomic factors, health behaviors, and mortality: Results from a nationally representative prospective study of US adults. *Journal of the American Medical Association, 279*(21), 1703–1708. doi:10.1001/jama.279.21.1703 PMID:9624022

Lara, M., Rosenbaum, S., Rachelefsky, G., Nicholas, W., Morton, S. C., Emont, S., & Weiss, K. B. et al. (2002). Improving childhood asthma outcomes in the United States: A blueprint for policy action. *Pediatrics, 109*(5), 919–930. doi:10.1542/peds.109.5.919 PMID:11986457

Maryland Department of the Environment (DOE). (2017). What is environmental justice? Baltimore, MD: DOE. Retrieved from http://www.mde.state.md.us/programs/CrossMedia/EnvironmentalJustice/WhatisEnvironmentalJustice/Pages/Programs/MultimediaPrograms/Environmental_Justice/ej_intro/index.aspx

McGovern, F. E. (1990). The Alabama DDT Settlement Fund. *Law and Contemporary Problems, 53*(4), 61–78. doi:10.2307/1191951

National Center for Health Statistics. (2013). *Health, United States, 2012: With special feature on emergency care*. Hyattsville, MD: NCHS.

National Center for Health Statistics. (2015). Health, United States, 2014. With special feature on adults aged 55-64 (Table 80).

National Wildlife Federation. (2017). What is biodiversity? Merrifield, VA; National Wildlife Federation. Retrieved from http://www.nwf.org/Wildlife/Wildlife-Conservation/Biodiversity.aspx

OECD. (2016). *Education at a Glance 2016. OECD Indicators*. Paris: OECD Publishing. doi:10.187/eag-2016-en

Pellow, D. N. (2002). *Garbage wars: the struggle for environmental justice in Chicago*. Cambridge, MA: MIT Press.

ReACTtoolkit.net. (2017). Resilience Adaptation Community Toolkit. Retrieved from http://reacttoolkit.net/

Rebitzke, J. (2004). Environmental Justice Case Study: DDT Contamination. Triana, AL. Ann Arbor, MI: University of Michigan. Retrieved from http://www.umich.edu/~snre492/triana.html

RoHS Guide Compliance. (2017). Available at: http://www.rohsguide.com/

Sebelius, K. (2012). Message from Secretary Kathleen Sebelius. In: PLAN-AT-A-GLANCE 2012. Environmental Justice Strategy and Implementation Plan. Washington, D.C.: U.S. HHS. Retrieved from https://www.hhs.gov/sites/default/files/environmentaljustice/plan_at_glance.pdf

Shea, J. (2007). The crisis in America's emergency rooms and what can be done. Washington, D.C.: The Heritage Foundation. Retrieved from http://www.heritage.org/health-care-reform/report/the-crisis-americas-emergency-rooms-and-what-can-be-done

Sommers, A.S.., Boukas, E.R., & Carrier, E. (2012). Dispelling myths about emergency department use: Majority of Medicaid visits are for urgent or more serious symptoms. *Res Brief,* July(23), 1-10.

Tang, N., Stein, J., Hsia, R. Y., Maselli, J. H., & Gonzales, R. (2010). Trends and characteristics of US emergency department visits. 19972007. *Journal of the American Medical Association*, *304*(6), 664–670. doi:10.1001/jama.2010.1112 PMID:20699458

United Nations. Framework Convention on Climate Change. (2014). Kyoto Protocol. New York: UN. Retrieved from http://unfccc.int/kyoto_protocol/items/2830.php

U.S. Department of Education. (2015). Office of Elementary and Secondary Education, Consolidated State Performance Report, 2013–14 (table 219.46). *Digest of Education Statistics.*

U.S. Department of Energy. (2017). DOE Announces Nearly $20 Million to Accelerate Energy- Efficient Transportation and Systems. Washington, D.C.: US DOE. Retrieved from https://energy.gov/eere/articles/doe-announces-nearly-20-million-accelerate-energy-efficient-transportation-and-systems

U.S. Department of Health and Human Services. (2012). PLAN-AT-A-GLANCE 2012 Environmental Justice Strategy and Implementation Plan. Retrieved from https://www.hhs.gov/sites/default/files/environmentaljustice/plan_at_glance.pdf

U.S. Environmental Protection Agency (EPA). (2016). DDT – A brief history and status. Retrieved from https://www.epa.gov/ingredients-used-pesticide-products/ddt-brief-history-and-status

U.S. Environmental Protection Agency (EPA). (2017a). Environmental justice. Retrieved from https://www.epa.gov/environmentaljustice

U.S. Environmental Protection Agency (EPA). (2017b). EJScreen: Environmental Justice Screening and Mapping Tool. Retrieved from https://www.epa.gov/ejscreen

World Wildlife Fund. (2017). EU Bird and Habitat Directives: What is at stake? Retrieved from http://www.wwf.eu/what_we_do/biodiversity/eu_birds_and_habitats_directives/

KEY TERMS AND DEFINITIONS

Biodiversity: The variety and abundance of species on our planet, as well as the genetic diversity within those species, and the diversity of entire habitats and ecosystems.

Electric Vehicle (EV): Also referred to as an electric drive vehicle, uses one or more electric motors or traction motors for propulsion; it may be powered through a collector system by electricity from off-vehicle sources, or may be self-contained with a battery or generator to convert fuel to electricity.

Environmental Justice: The fair treatment and meaningful involvement of all people regardless of race, color, national origin, or income, with respect to the development, implementation, and enforcement of environmental laws, regulations, and policies.

European Union (EU): An economic and political union between 28 European countries founded in the aftermath of World War II with headquarters in Brussels, Belgium; today its areas of interest have expanded to include climate, environment, health, external relations and security, and migration.

Gender Gap: A disproportionate difference, or disparity, in opportunities, status, or attitudes between men and women.

Geocoding: Using a technology tool, GIS software, to locate regions or specific neighborhoods worldwide (see GIS below).

Geomapping: Using a technology tool, GIS software, to visually depict selected indicators on a map.

GIS: Geographic Information Systems (GIS), a technology tool used to store, retrieve, manage, display, and analyze all types of geographic and spatial data.

Health Disparities: Preventable differences in the burden of disease, injury, violence, or opportunities to achieve optimal health that are experienced by socially disadvantaged populations.

Healthy People 2020: A national blueprint for American health that provides science-based, 10-year national objectives to help Americans live high-quality, longer lives free of preventable disease, disability, injury, and premature death.

Illiteracy: The inability to read and write.

Literacy: The ability to read and write.

MOOC: Massive Online Open Courses, offered for free on the Internet by universities, technology company, and other organizations.

Pandemic: An epidemic occurring or over a very wide geographic area and affecting an exceptionally high proportion of the population.

Sustainability: The ability to meet the needs of the present without compromising the ability of future generations to meet their own needs; the capacity to improve the quality of human life while living within the carrying capacity of the Earth's supporting eco-systems.

Chapter 5
DOI Theoretical Framework:
Adopting Innovative Technologies

ABSTRACT

Change is not easy! People adhere to old routines and habits tenaciously. Most people are slow to accept new ideas, new products, in short, innovations. When it comes to new technologies that can aid in adaptation to climate change, there is fierce resistance from farmers (to sustainable agriculture), from the fossil fuels industries (to sustainable energy), from developers (to going green), and the list goes on. While a new technology does involve a certain investment of time and money at first, it is cost effective and profitable in the long term. When it comes to sustainability, nothing less than the future of our planet is at stake, so it is incumbent upon us to find a way to "sell" the innovations to the masses. The Diffusion of Innovations (DOI) Theoretical Framework provides an effective, structured means of doing this; its efficacy has been established for hundreds of innovations, and it is particularly suited to technologies.

INTRODUCTION

DOI Theory was the brainchild of Everett M. Rogers, who was born on his family's farm in Carrol, Iowa in 1931. As a child, Rogers observed the reluctance of his father to adopt a new hybrid seed corn, even though it was drought-resistant and yielded 25 percent more crop. Drought hit the farm in 1936 and the hybrid seed corn in a neighboring farm survived while the corn

DOI: 10.4018/978-1-5225-3414-3.ch005

on the Rogers farm wilted. This experience made an indelible impression on the young Everett Rogers. He wondered why some farmers adopted innovations while other resisted innovations. Rogers enrolled in Iowa State University and studied rural sociology and the diffusion of agricultural innovations (Rogers, 2003). Out of his early work in diffusion of agricultural innovations arose an interest in a general theory about adoption of innovations. He developed a generalized diffusion theory, and posited that his theory explained social change. Today DOI Theory is widely used around the globe for a variety of fields. It is the theoretical framework of choice for the diffusion of new technologies in adaptation to climate change.

DIFFUSION OF INNOVATIONS THEORY

Everett M. Rogers seminal work, *Diffusion of Innovations Theory*, was first published in 1962. He continued to update and publish revised editions (Rogers, 1983, 1995, 2003) until a year before he died. By the time he wrote his fifth and final edition of *Diffusion of Innovations* (2003), DOI theory had become so widely accepted as a valuable framework for social change that Rogers estimated that there were some 5,200 publications about *DOI* Theory.

Rogers (2003, pp. 11-12) defined innovation as "an idea, practice, or object that is perceived as new by an individual or other unit of adoption" and diffusion as "the process in which an innovation is communicated through certain channels over time among the members of a social system." Following are the four main elements in DOI Theory:

- Innovation;
- Communication channels;
- Time;
- The social system.

Innovation

Rogers noted that the perceived characteristics of an innovation influences its rate of adoption. He (2003) listed five characteristics as the most predictive of the adoption rate:

- Relative advantage (the degree to which the innovation is perceived to be an improvement over the status quo);
- Compatibility (the degree to which the innovation is perceived to be consistent with target audience's existing values, experiences, and needs);
- Complexity (the degree to which the innovation is perceived to be difficult to understand and use);
- Trialability (the degree to which the innovation can be experimented with on a limited basis);
- Observability (the degree to which the results of the innovation are visible to others).

The first two characteristics—relative advantage and compatibility—are the most important. The simpler an innovation appears to be to the target audience, the more likely they are to adopt it. Another characteristic Rogers (2003) cites is the ability to re-invent the innovation, that is, to customize it and make it one's own. This ability to re-invent an innovation makes it more attractive to the target audience (Rogers, 2003).

Communication Channel

Marshall McLuhan famously stated back in 1964, "The medium is the message." He posited that the medium itself imbedded its meaning in the message. Communication channels are media through which innovations are communicated. Mass media channels such as radio, television, and newspapers are effective at reaching large numbers of people. Interpersonal channels consist of face-to-face exchanges between people. The Internet provides instantaneous interactive communication around the world (Rogers, 2003).

Time

Rogers coined the phrase "innovation decision process" to refer to the amount of time it takes from the introduction of an innovation to its adoption. This process generally follows a time-ordered sequence of five stages, as shown in Table 1.

It was clear to Rogers even as a child that people differ in their willingness to adopt innovations: some adopt innovations fairly readily, while other take

Table 1. Stages in the innovative decision process

Knowledge	First Exposed to Innovation
Persuasion	Actively seeks information about innovation
Decision	Weighs pros/cons of adopting innovation and makes a decision
Implementation	Employs the innovation and forms an opinion of its usefulness
Confirmation	Finalizes decision to adopt the innovation

Source: Rogers (2003).

years to adopt an innovation, even when there is evidence of its benefits to others. Adoption may be by an individual or an organization.

The Transtheoretical Model (TTM) (Prochaska et al., 1992, 1994, 2009; DiClemente & Prochaska, 1999), one of the most widely used models for behavior change, works well in conjunction with Rogers' DOI theoretical framework. The TTM is grounded in the psychological concept of self-efficacy (Bandura, 1982, 1989), which refers to the degree of confidence individuals have in their ability to make a desired behavior change. If a person does not have confidence in his or ability to change (e.g., to adopt a technology or implement a new strategy), it is highly unlikely that he or she will make the change. Positive peer role models are useful in boosting people's self-confidence; once they see someone else similar to them do a task, it becomes less intimidating. For example, teaching senior citizens to use computers often meets resistance (I'm too old; I don't need it; It's too hard); however, the seniors may be motivated to learn by the desire to write a life story or to see photos of their grandchildren (Duay & Bryan, 2008). Their confidence can be boosted by interacting with other seniors who have learned to use the Internet recently. The Internet can be a great tool for seniors who have limited mobility, and can help them to locate health information so that they are informed healthcare consumers.

TTM posits that behavior change is a gradual process that moves through five Stages of Change. The first stage is precontemplation, when it becomes clear to others around the individual that he/she needs to change but he/she is not yet ready to do so; Motivational Interviewing (MI) (Miller & Rollnick, 2013) may be used to make the individual aware of his or her problem. After the individual recognizes the need to change and feels motivated to change, he or she reaches the second stage of the TTM, called contemplation; at this stage, the individual is thinking seriously about making a change. The third stage is preparation, when the individual intends to take action in the immediate future and actively plans for the change. The fourth stage is action, when the individual changes his/her behavior; and the fifth stage is maintenance, when the individual consolidates and maintains the behavior change. Maintenance continues indefinitely, with consolidation and adoption of the new behavior. Table 2 shows the correspondence between the stages of DOI theory and TTM Stages of Change.

Social System

Rogers (2003, p. 22) defined "innovativeness" as "the degree to which an individual or other unit of adoption is relatively earlier in adopting new ideas than the other members of the system." He classified members of a social system on the basis of their innovativeness, that is, how long it took them to adopt an innovation; Rogers's classified adopters into five categories, which are shown in Table 3.

The first to adopt an innovation are innovators; they are active information seekers who can handle higher levels of uncertainty than can the other categories of adopters. Innovators tend to have more mass media exposure and more extensive interpersonal networks than do the other categories of adopters. DOI occurs within the context of a social system, and the social norms, or established behavior patterns, impact the rate of adoption of an

Table 2. Correspondence of doi theory & TTM stages of change

Diffusion of Innovation Theory Stage	Transtheoretical Model Stages of Change
Knowledge	Precontemplation
Persuasion	Contemplation
Decision	Preparation
Implementation	Action
Confirmation	Maintenance

Source: Rogers (2003).

Table 3. Categories of adopters

Innovators	Quickly finalize their decision to adopt the innovation
Early adopters	Respected opinion leaders, willing to try new ideas
Early majority	Careful, thoughtful, but accepting change more quickly than the average
Late majority	Skeptics who will try something new only after majority does
Laggards	Traditionalists critical of new ideas, will accept change only after it is mainstream

Source: Rogers (2003).

innovation. Opinion leaders play a key role in the adoption or rejection of an innovation because they influence the perceptions of many others in the social system. Rogers (2003) noted the importance of understanding the respective roles of early adopters and later adopters in creating a tipping point in an organization for adopting an innovation. A comparison between early adopters and later adopters in terms of socioeconomic status, personality values, and communication behaviors is shown in Table 4.

Early adopters tend to embrace new products, services and technology before most other people do. They want to be the first to buy a 3D HDTV or to get a brand-new app for their phone. They enjoy technology and are eager to try new products. If you want to identify the early adopters in a group, ask them about their attitude toward new products compared to others; have them select one of the following statements that best describes how they perceive themselves (Gendelman, 2017):

- I usually am the first to try a new product;
- I usually am among the first to try a new product;

Table 4. Early adopters vs. later adopters

Early Adopters	Late Adopters
1. Socioeconomic Status	
More highly educated Higher literacy rates Higher social status (income, lifestyle, occupational prestige, self-perceived identification with a social class) More upwardly mobile Part of larger-sized units (farms, schools, companies)	Less highly educated Lower literacy rates Lower social status (income, lifestyle, occupational prestige, self-perceived identification with a social class) Less upwardly mobile Part of smaller-sized units (farms, schools, companies)
2. Personality Variables	
More empathic Less dogmatic Greater ability to deal with abstractions Greater rationality More intelligent More favorable attitude toward change Better able to handle uncertainty and risk More favorable attitude toward science Less fatalistic Higher aspirations	Less empathic More dogmatic Lesser ability to deal with abstractions Lesser rationality Less intelligent Less favorable attitude toward change Less able to handle uncertainty and risk Less favorable attitude toward science More fatalistic Lower aspirations
3. Communication Behavior	
Greater social participation Greater interpersonal connectedness More widely travelled and cosmopolite	Lesser social participation Lesser interpersonal connectedness Less widely travelled and cosmopolite

Source: Rogers (2003).

- I am usually in the middle when it comes to trying a new product;
- I am usually one of the last to try a new product;
- I usually am the last to try a new product.

In addition to being open to trying new products, the early adopters should be opinion leaders who want to influence others and who feel passionate about the product/cause. It is important to make the early adopter feel special, so that he or she will feel that their opinions are held in high regard.

Figure 1 shows Rogers" (2003) bell curve for the adoption of an innovation for the five categories of adopters.

Rogers' bell curve is not symmetrical; three adopter categories fall to the left of the mean and only two to the right. Rogers considered breaking laggards into two categories, but decided against it because "laggards seem to form a fairly homogeneous category" (Rogers, 2003, p. 281). In a normal bell curve, 68 percent of the individuals fall within 1 standard deviation of the mean, and 95 percent fall within two standard deviations from the mean. In Rogers' bell curve, the early and late majority comprise 68 percent, the early adopters add 13.5 percent, and the laggards take up the whole right tail

Figure 1. Categories of innovativeness
Source: Rodgers (1995)

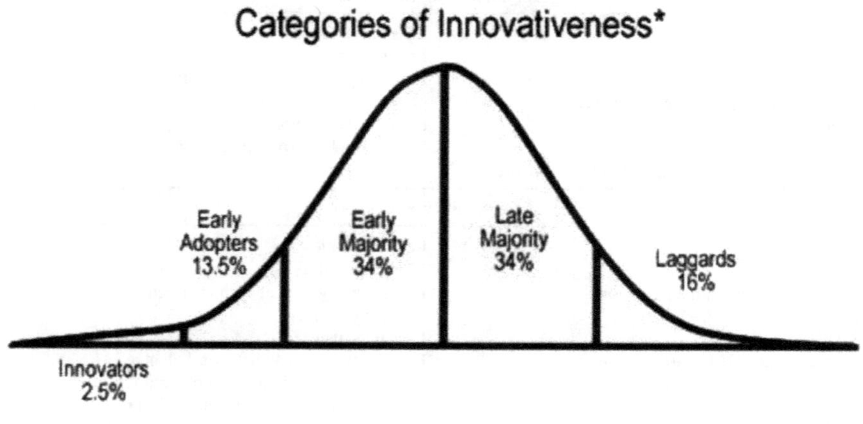

of the curve at 16 percent. The innovators comprise a slim 2.5 percent. This bell curve shows just how hard it is to get some people to change!

CONTINUUM OF HEALTHY BEHAVIOR CHANGE

Convincing individuals (or organizations) to change their behavior begins with education about the consequences of adopting the innovation; after that a social marketing campaign is needed to persuade people about the relative advantages of the innovation vis-a-vis the status quo (Rogers, 2003); the third step is to implement policy changes that make the environment supportive of the innovation. Initiating healthy behavior change is a time-consuming process that follows a continuum, shown in Figure 2.

The anti-smoking campaign in the United States provides an excellent case example of how the healthy behavior change continuum works in practice. It is worth taking a closer look at how this very intense and long-term effort finally succeeded in reaching a tipping point where nonsmokers, not smokers, were in the driver's seat and where laws were passed that made public transportation, workplaces, schools, restaurants, and other public places *tobacco free*. The tobacco companies in the United States, who had powerful lobbies in Washington, D.C., put up an intense resistance and poured vast amounts of money into fighting the public health campaign. Nevertheless, the persistence of the national effort eventually won over the public. This case example demonstrates what Rogers observed many years ago, namely, that most people do not readily change their lifestyle habits, even when there is clear evidence that it is in their best interest (Rogers, 1962; 1983, 1995, 2003).

Case Example 1: Anti-Smoking Campaign in the United States

It all began on January 11, 1964, when Luther L. Terry, MD, Surgeon General of the U.S. Public Health Service, announced to the American public that

Figure 2. Continuum of healthy behavior change

smoking tobacco was hazardous to one's health, that it caused lung cancer, laryngeal cancer, and chronic bronchitis. His announcement was based on a report by the Surgeon General's Advisory Committee on Smoking and Health; this report was the result of research that included more than 7,000 articles on tobacco smoke and disease. The U.S. Surgeon General's warning did not make a big impact at first; the public response was tepid (Centers for Disease Control and Prevention, 2009). Why didn't all the tobacco smokers take heed and quit smoking? There are two main reasons for their low response: first, nicotine is one of the most addictive substances on earth, 6-8 times more addictive than alcohol, on a par with heroin; this makes smoking cessation very difficult and uncomfortable, with extreme cravings, weight gain, and other side effects. Second, smoking cigarettes and cigars was a social behavior that gave smokers great satisfaction; it was a habit deeply ingrained in their lives. Changing such a behavior is a complex process that takes time. The acceptance and adoption of a new behavior requires a change in public perception, which is facilitated by an effective health education campaign (Oldenburg & Parcel, 2002). The DOI Theoretical Framework provides a roadmap for promoting and gaining widespread acceptance of an innovation, including a new behavior.

In the aftermath of the 1964 report by U.S. Surgeon General Luther L. Terry, MD, the government required that cigarette packages add warnings about the hazards to health posed by smoking tobacco. Advertisements for cigarettes continued on radio, television, movies, and print media with icons such as Joe Camel and Marlboro Man. The image associated with smoking cigarettes was cool, sexy, and hip. The jingles for the commercials were catchy—even children picked up the tunes and repeated them, and their parents could purchase candy cigarettes for them to "smoke." In 1969, the U.S. Surgeon General issued another report stating that low birthweight was associated with tobacco use. Bowing to pressure, Congress passed the Public Health Cigarette Smoking Act, banning cigarette commercials from radio and TV, effective January 2, 1971. Cigarette advertising continued in print sources such as magazines, and product placements of brand name cigarettes in movies glamorized smoking. Additional federal restrictions were passed, banning smoking on airplanes and cigarette advertising on billboards.

The next step in tobacco control was to take DOI to the state level. Individual states created tobacco-free areas, then tobacco-free buildings, and then tobacco-free campuses and workplaces. Slowly, a tipping point was reached, and the onus was on the *smoker* to leave and smoke outside. Of course, now workers who do not smoke complain that smokers get "too many breaks" at work, because they have to go outside! But there is big health benefit--employees are not exposed to secondhand smoke.

Recently, the public health campaign *Tips from Former Smokers* showed people who lost limbs and died from cancer caused by smoking. *Tips from Former Smokers* was a multilevel national government-funded anti-smoking campaign launched by the Centers for Disease Control and Promotion in 2012 (McAfee et al., 2013). This anti-smoking campaign showed graphic videos and photos of dying people. The message proved to be quite effective. It ran on TV stations, aired on the radio, and was advertised on the Internet, in print, and on billboards. A survey was sent to adult smokers and non-smokers before the *Tips* campaign and after 12 weeks of exposure to the campaign. *Tips* reached nearly 80 percent of U.S. smokers and was associated with a 12 percent relative increase in quit attempts and a 3.7 percent absolute increase in quit attempts. The ultimate goal of the national anti-smoking campaign is to get all smokers to stop and to prevent young people from starting. Tobacco use is the leading preventable cause of death in the United States. *Healthy People 2020* made *not using tobacco* a leading national health indicators for adolescents and adults in the United States (*Healthy People 2020*, 2017). Another leading health indicator is environmental quality, which targets air quality, including exposure to secondhand smoke.

Corporations can partner with governments, schools, and department of health to work toward achieving health goals. For example, the drugstore chain CVS made a strong anti-smoking statement when it stopped selling cigarettes and other tobacco products. Otis Brawley, Chief Medical Officer for the American Cancer Society, called CVS's decision an "an act of corporate courage." He noted that the American Cancer Society has been trying to convince pharmacies to ban cigarettes for several years. Convenience and availability are key factors affecting smokers. Brawley noted that "Studies show that being forced to travel just two extra blocks can deter someone from buying cigarettes" (Szabo, 2014).

Smoking cessation programs are offered in many workplaces and schools. However, the decision to stop smoking is an individual one. Such a difficult decision takes time, as the individual passes through the TTM Stages of Change. The first stage, precontemplation, is where many addicts languish for decades or longer, not able to take action. Their families and friends know that they have a problem, but they themselves do not acknowledge it. The addict is in denial and rationalizes his or her reasons for continuing to smoke. e.g., "I can handle it, it's not that many cigarettes, it keep my weight down, etc.). Once a smoker comes to the realization that he or she has a problem and is addicted to cigarettes, and wants to change, the smoker has reached the stage of contemplation. It is generally accepted that the stage of contemplation is reached when the person making the change decides to do so within six months. Next, during the stage of preparation the smoker plans his or her smoking cessation. What aids will he or she need (nicotine patch or gum, Chantix or other medication) to help deal with cravings and other withdrawal symptoms? What times of the day (e.g., meals) and situations (e.g., attending parties, getting together with friends or family members who smoke) are the highest risk for relapsing? The smoking cessation counselor helps the smoker seeking to quit to identify how to avoid high-risk situations or people that make it likely that he or she will smoke a cigarette. After preparation comes the action stage, when the smoker stops smoking cigarettes. Smokers who have just quit need support and patience from those around them at home, at work and at play. The final TTM Stage of Change is maintenance; the person has changed and maintains the new healthy behavior.

The following case examples demonstrate how the DOI theoretical framework can be used for emergency management, to assess how best to dispense vital information for emergency preparedness to school systems, corporations, universities, municipalities, and health departments, and to facilitate diffusion and adoption of technologies during emergencies. For DOI to be effective, adequate resources have to be allocated to purchase and deploy the needed technological innovations (e.g., maritime satellite phones in case example 2).

Case Examples 2 and 3: A Comparison of Disaster Relief After Earthquakes in Sichuan, China (2008) and East Japan (2011)

At 2:28pm on May 12, 2008, when most adults were at work and children were in school, a massive earthquake occurred in Sichuan (formerly known as Szechwan), a mountainous and relatively undeveloped province of southwest China; its epicenter was 50 kilometers (31 miles) from Chenchen, the capital of Sichuan (Zhou et al., 2012). The fatalities were immense—an estimated 70,000 people (Aiming, 2009) died. In the aftermath of the disaster, researchers (Zhou et al., 2012) conducted focus group DOI sessions with health professionals who had worked in the earthquake rescue and relief effort. The country's emergency response network was completely inadequate to deal with the earthquake, and rescue workers reported that there were shortages of maritime satellite cell phones, large tents for medical use, and fully equipped ambulances. The lack of cell phones made diffusion of health information and services by rescue health workers impossible in remote areas, and the lack of adequate medical transport resulted in delays in getting medical services which, in turn, caused fatalities to skyrocket. Rainy weather and local languages posed additional major challenges in disseminating services after the disaster (Zhou et al., 2012).

At 2:46 pm on March 11, 2011, almost the identical time of day as the Sichuan earthquake, a massive 9.0 earthquake hit the Tohoku region of Honshu in Eastern Japan, a relatively remote region with a harsh climate. The earthquake was the most severe to ever hit Japan and was followed by a tsunami that caused damage to reactors at the Fukushima Daiichi Nuclear Power Plant, which led to the evacuation of hundreds of thousands of residents. The total fatalities were estimated at 15,660 (Japanese National Police Agency, 2011)—a huge loss, but nowhere near the Sichuan fatalities of 70,000. A Chinese study (Otani, 2012) compared the two earthquakes and concluded that Japan had the following technologies in place whereas China did not: a national earthquake emergency plan, fixed evacuation sites with experienced staff, and earthquake education, which educated the public about what to do in case of an earthquake. Both Japan and China lacked adequate supplies, but that may have been due to the unprecedented scope of the earthquakes. With climate change continuing to cause more frequent and severe natural disasters around the world, it behooves all nations, states and provinces, cities, and communities to prepare as follows (Otani et al., 2012):

- Formulate detailed national emergency plans and centralize control of disaster relief efforts.
- Arrange for fixed evacuation sites and make these locations and safe routes to them known to public through signage, PSAs, Internet, and other media.
- Arrange emergency transportation to evacuate people from impacted area to fixed evacuation sites.
- Train rescue workers on an ongoing basis--online modules are useful (see Weiss, 2016).
- Provide natural disaster education to all ages, from school children through senior citizens, using a variety of media, including books, posters/fact sheets that can be posted on refrigerators, flyers, newspapers, billboards, television, Internet and billboards.
- Stockpile adequate supplies, including well-equipped ambulances.

- Ensure that all communication channels have backup generators in the case of power outages.
- Ensure that all information dispensed is in languages that the community can understand and read, and geared toward age and literacy level of community.
- Use DOI theoretical framework to dispense materials in school systems, corporations, universities, municipalities, and health departments to ensure that it is understood and adopted by the public.

In addition, in regions where earthquakes and tsunamis occur, mandatory drills should be held to train residents how to minimize risks from earthquake-induced secondary effects such as tsunami. Think about it—in the United States, all cruise ships have mandatory safety drills before going out in good weather on calm seas to resort destinations. In regions at high-risk of natural disasters, such drills could save thousands of lives.

Another difference between Japan and Chinese is their classification: Japan is a "developed country" while China is a "developing country." Japan has the highest life expectancy at birth—83.5 years—of any nation, and is listed 20th on the list of nations ranked by Human Development Index (HDI), which measures well-being along three dimensions of human development: health, education and income (United Nations Development Programme, 2015a).

Over the past four decades, China has experienced unprecedented economic and social growth rates, with Gross Domestic Product (GDP) increasing nearly 10 percent per year. China reached all the Millennium Development Goals (MDGs) by 2015. However, with a population of 1.3 billion, China is still classified a developing country because of its low per capita income and large numbers of poor, especially in rural areas (World Bank, 2016). China's rapidly accelerating urbanization is approaching a critical stage, with 310 million more people projected to live in cities by 2030; these burgeoning cities would account for 70 percent of the total population (United Nations Development Programme, 2015a).

China's rapid development has come at a high cost to the environment; China is the single largest emitter of carbon dioxide, spewing more than 10.5 million Ktons into the air in 2014. Beijing is one of the most polluted cities on the planet; The United States Embassy in Beijing monitors daily air quality for fine particulate matter, known as PM2.5, which can be inhaled into the lungs, causing respiratory distress, especially in children, pregnant women, the elderly, and individuals with allergies, asthma and other respiratory illnesses. This PM 2.5 particulate matter has been at an average of 100 micrograms per cubic meter since 2008, about six times the level deemed safe by the U.S. Environmental Protection Agency (Kuo, 2014)! DOI could be used to assess how to dispense safety information on outdoor air quality to vulnerable populations in China, so that they can remain indoors as much as possible and wear protective masks when going outside. In addition, DOI could be used in a public health campaign to advocate for stricter pollution controls and adherence to the regulations that already exist.

DIFFUSION OF INNOVATIONS FOR NEW TECHNOLOGY

Leaders who plan to use DOI to market a new technology should use audience segmentation to market the innovation, altering the message to suit the distinctive characteristics of each adopter group. For example, the Internet may be the communication channel of choice for early adopters, while a face-to-face information session or printed flyer may work better for other adopter categories. In addition, buyers will be more likely to have a positive attitude toward change if they feel that their input is valued, that they are part of the process (focus groups, market surveys). Leaders should include the early adopters in the process of persuading the late adopters to change (Stichler, 2011).

Returning to our Continuum of Healthy Behavior Change (Figure 2), the first step in introducing a new technology is education--making consumers aware of and knowledgeable about it, in language that is clear, understandable and respectful of gender, ethnic and racial background, age, and sexual orientation. Reading level is important—leaders have to know their audience's comprehension level and grasp of the written language; media such as videos and webinars can be quite effective. Cultural competency is equally important—leaders should familiarize themselves with the cultural customs of their target audiences so as to appeal to their tastes and preferences and avoid inadvertent misunderstandings.

The second step is to promote the new product, idea or behavior through a social marketing campaign to change the audience's attitude and thereby increase the acceptability of the innovation (Simons-Morton et al., 1995). It is important to identify the early adopters and focus on getting them on board, as they will help to influence the later adopters. How can early adopters be identified? They are the usually the first to adopt new technologies, such as the Iphone, Ipad, 3D LED HDTV, etc. A German study sought to identify the early adopters who would be willing to buy electric vehicles (EVs)

(Plotz et al., 2014). The study looked primarily at Battery Electric Vehicles (BEVs) but also at Plug-in Hybrid Electric Vehicles (PHEVs). PHEVs reduce dependence on gasoline and reduce emissions, while BEVs use no gasoline and have zero emissions. In addition to being environmentally friendly, reducing auto emissions, and decreasing reliance on fossil fuel, EVs are quieter and get better mileage than their traditional counterparts. For example, the 2012 EV Civic gets an average of 44 mpg versus 32 mpg for the conventional Civic (Park, 2014). The economics of a car—its investment and operating costs—influence purchasers' decisions. The impact of sticker price, fuel prices, and fuel efficiency on buying decisions is well established (Trainer & Winston, 2007; Sanini & Vyas, 2005). In Germany, the more fuel-efficient diesel vehicles show more kilometres travelled (22,300) than do vehicles with gasoline engines 11,800) (Follmer et al., 2010). The disadvantages of EVs are that they tend to cost more to purchase than Internal Combustion Engine Vehicles (ICEVs), their range is more limited than that of ICEVs, and it takes time to charge them. Then there is the intangible association of luxury and comfort with ICEVs such as Mercedes and BMWs; of course, newer luxury EVs are being developed (e.g., Tesla in U.S.) that have the cachet of the old-school lux ICEVS (Park, 2014).

There are relatively few studies of EV buying habits in Germany. EVs make up a tiny fraction of the car market there. Yet Germany is the fourth largest producer of automobiles and auto sales are the backbone of its economy, so identifying and capturing an emerging EV market there has great money-making potential—as well as being good for the environment (Plotz et al., 2014)! Research has shown that, overall, drivers find EVs easy to drive (Dutschke et al., 2012; Peters at el., 2011), but even individuals who have driven EVs and evaluated them positively show only a modest interest in buying an EV (Dutschke et al., 2012; Graham-Rowe et al., 2012).

A UK study (Anable et al., 2011) found that the characteristics of early adopters were different for BEVs and PHEVs; they found that BEVs are considered for second household cars, whereas PHEVs were purchased as the main or sole vehicle. Low range anxiety and an EV-friendly environment were found to be the strongest drivers of EV adoption, followed by openness to EV technology and a high willingness to pay (Anable et al., 2011). Due to the high prices of EVs, it is extremely likely that early buyers will have above-average socioeconomic status.

What other characteristics do early adopters share? Rogers (2003) described early adopters as typically younger, with higher social status, more education, and more social connections than late adopters. Studies (Curtin et al., 2009;

Peters et al., 2011; Ozaki & Sevastyanova, 2011) have found that attitudinal variables such as environmental awareness, technology affinity, and the symbolic meaning of EVs are also characteristics associated with EV buyers (early adopters).

The German study (Plotz et al., 2014) analyzed socioeconomic variables such as gender, age, education, income, household size, and location, as well as attitudinal variables such as environmental awareness, technical affinity, and the symbolic meaning of EVs. Data were collected for 210 potential EV buyers who had expressed high interest through an online survey and for 969 EV owners; in addition, 27,918 driving profiles from a public survey of driving behavior (Mobility in Germany) (MiD, 2002) were analyzed (Plotz et al., 2014).

The EV owners had the following profile: they were 81.4 percent male, mean age 40.9 years, mean household size 2.5 persons, mean household income EURO 2001-3000 (categorical scale), 1.4 cars per household, 51 percent had a Master's degree or higher, and 42 percent worked in technical fields. Potential EV buyers with high interest were recruited through a hyperlink to the early adopter survey that was disseminated in online forums on electric mobility. The profile of the potential buyers with high interest was 91 percent male, aged 41-50, with an average of 2 cars per household. The driving profiles were categorized into groups based on whether an EV would be cost-efficient based on where the individual lived, his or her employment status and length of commute. EVs are cost-effective if the owner drives a sufficient number of kilometers (or miles) per year. The researchers identified driver profiles who would benefit economically from an EV purchase as full-time workers living in small to medium-sized cities. The projection was consistent with the results, which showed that actual EV users and those who expressed high interest in EV ownership were more likely to be male, work full-time, and live in small to medium-sized municipalities (Plotz et al., 2014). EVs are less likely to be purchased by residents of large cities because in Germany only a small percentage of large city dwellers own cars, and those that do own a car generally do not drive it enough to make investing in an EV cost-effective (Plotz et al., 2014).

Since 2014, when the German study (Plotz et al., 2014) of identifying early adopters and marketing EVs was published, the campaign to go electric in Germany has taken off in a big way as laws were passed to incentivize developing EVs and giving price breaks to consumers who purchase them. In

2016, Germany joined the International Zero-Emission Alliance and launched an incentive and investment program to accelerate the development of EVs; it includes a 4,000 euro discount for all-electric vehicle purchases. Furthermore, Germany has mandated that all vehicles registered in the country will be electric (emissions free) by 2030; this mandate is part of Germany's pledge to reduce carbon dioxide emissions by at least 80 percent by 2050 (Lambert, 2016). Rogers (2003) would most likely advise the German automobile manufacturers to create EVs that appeal to women, since the vast majority of the EV buyers and high-interest potential purchasers have been male!

In sum, healthy behavior change moves along a continuum from education (knowledge), to social marketing, to laws supporting the innovation. In our case example, the U.S. Surgeon General's 1964 report on smoking informed the public about the dangers of tobacco (education); subsequently the anti-smoking message was diffused around the country using various communication channels (e.g., PSAs, warnings on packages) and messages (social marketing); next, laws were passed banning cigarette commercials on TV, banning smoking on airplanes, and eventually creating tobacco-free campuses and workplaces. The anti-smoking movement has made tremendous progress, but it took a long time because of the resistance by the tobacco lobby as well as the profound attitudinal change required to adopt a new social norm enforced through practice and policy. After all, smoking had been glamorized in the movies for a century, and on TV for more than 50 years! The tipping point has been reached, and there is no going back. No smoking has become the "new normal"!

The movement to replace conventional vehicles with EVs is still in the early stages compared to the smoking cessation campaign. However, the Kyoto Protocol and the 2015 Paris Climate Accord (United Nations, 2015b) have mandated targeted reductions in CO_2, emissions, which has pushed European countries toward the third step in the continuum of healthy behavior change—i.e., laws to support the development and adoption of EVs—in a relatively brief period of time. Norway paved the way, with laws providing generous incentives for EV purchases. The Netherlands, Sweden, Germany and other countries are following suit. Timing is crucial; passing laws alone will not work until the attitudes have changed; witness what happened during Prohibition! Speakeasies, bootleg liquor and the rise of the Mafia were a few of the unwanted byproducts of a law that was passed without proper education or a change in attitudes toward alcohol. Social marketing campaigns are needed to persuade different market segments that EVs are the best choice for them. The economic incentives alone will hook individuals who are more flexible and willing to adopt innovative technologies (the innovators and early adopters, in DOI's terms), but to convince the majority, social marketing campaigns need to be mounted on multiple communication channels (e.g., print media, radio, TV, Internet, and social media) (See Chapter 6 for a discussion of EV market penetration in Norway.)

CONCLUSION

Successful adaptation to climate change requires a profound change in attitudes among individuals, government leaders, corporative CEOS who set policy, and civic leaders. First, there has to be acknowledgment that there is a problem—that, in fact, temperatures and sea levels are rising, causing severe natural disasters, such as the 2008 earthquake in Sichuan, China and the 2011 earthquake and tsunami in Tohuku, Japan. There are still a number of world leaders who are in denial about climate change, but the Paris Climate Conference in 2015 (United Nations, 2015b) showed that the world has reached a tipping point in acknowledging that the environmental degradation and other impacts of climate change needed to be addressed with a concerted, long-term global effort.

DOI Theory provides a useful framework for the spread and adoption of innovations such as EVs by society. It also is useful in disaster relief planning for earthquakes and hurricanes, and emergency preparedness for climate-change-related events such as flooding caused by storm surges and severe heat waves.

Facilitating the adoption of a new technology by the public requires persistence, and time, as people move through a continuum of behavior change that begins with education, moves on to social marketing, and concludes with the passage of laws and policies that reinforce adoption of the technology.

REFERENCES

Aiming, L. (2009). The mechanism and hazard produced by the 2008 Mw8.0 Sichuan earthquake, China. *The Almanac of China*, *40*. (in Japanese).

Anable, J., Schuitema, G., Skippon, S., & Kinnear, N. (2011). *Who will adopt electric vehicles? A segmentation approach of UK consumers. Proceedings from ECEE 2011 Summer Study*. France: Belambra Presqu'lle de Giens.

Bandura, A. (1982). Self-efficacy mechanism in human agency. *The American Psychologist*, *37*(2), 122–147. doi:10.1037/0003-066X.37.2.122

Bandura, A. (1989). Human agency in social cognitive theory. *The American Psychologist*, *44*(9), 1175–1184. doi:10.1037/0003-066X.44.9.1175 PMID:2782727

Centers for Disease Control and Prevention (CDC). (2009). Smoking and tobacco use: history of the Surgeon General's report on Smoking and Health. Retrieved from https://www.cdc.gov/tobacco/data_statistics/sgr/history/index.htm

Curtin, R., Shrago, Y., & Mikkelson, J. (2009). Plug-in Hybrid Electric Vehicles. University of Michigan, Ann Arbor, MI. Retrieved from http://www.ns.umich.edu/Releases/2009/Oct09/PHEV_Curtin.pdf

DiClemente, C. C., & Prochaska, J. O. (1999). Toward a Comprehensive Transtheoretical Model of change: Stages of change and Addictive Behaviors. In W. R. Miller & N. Health (Eds.), *Treating Addictive Behaviors* (2nd ed.). New York, London: Plenum Press.

Duay, D. L., & Bryan, V. C. (2008). Seniors and computers: A love/hate relationship. Paper for American Education Research Association (AERA) 2008 SIG-Adulthood and Aging Special Interest Group. New York: NY: International AERA (paper session).

Dutschke, E., Schneider, U., Sauer, A., Wietschel, M., Hoffman, J., & Domke, S. (2012). Roadmap sur Kundenakzeptanz. Zntrale Ergebnisse der sozialwissenschaftlichen Begleitforschung in den Modellregionen Herausgeber: Bendesministerium fur Verkehr, Bau und Stadtentwicklung (BMVBS). Berline: Karlsruhe.

Gendelman, V. (2017). Identifying and targeting early adopters. Keego Harbor, MI: Printwand. Retrieved from http://www.printwand.com/blog/identifying-and-targeting-early-adopters

Graham-Roe, E., Gardner, B., Abraham, C., Skippon, S., Dittmar, H., Hutchins, R., & Stannard, J. (2012). Mainstream consumers driving plug-in battery-electric and plug-in hybrid electric cars: A qualitative analysis of responses and evaluations. *Transportation Research Part A, Policy and Practice*, *46*(1), 140–153. doi:10.1016/j.tra.2011.09.008

Healthy People 2020. (2017). Social determinants of health. Washington, D.C.: U.S. Department of Health and Human Resources. Retrieved from https://www.healthypeople.gov/2020/about/History-and-Development-of-Healthy-People

Japanese National Police Agency. (2011). The Great East Japan Earthquake Information. Retrieved 2011.12.28 from http://www.jst.go.jp/pr/pdf/great_east_japan_earthquake.pdf

Kuo, L. (2014). Six years of Beijing air pollution summed up in one scary chart. U.S. Embassy, U.S. Department of State. Retrieved from https://qz.com/197786/six-years-of-bejing-air-pollution-summed-up-in-one-scary-chart/

Lambert, F. (2016). All new cars mandated to be electric in Germany by 2030. Cupertino, CA: Electrek. Retrieved from https://electrek.co

McAfee, T., Davis, K. C., Alexander, R. L., Pechacek, T. F., & Bunnel, R. (2013). Effect of the first federally funded anti-smoking national media campaign. *Lancet*, *382*(9909), 2003–2011. doi:10.1016/S0140-6736(13)61686-4 PMID:24029166

McLuhan, M. (1964). *Understanding Media: The Extensions of Man. New York: Mentor Press; reissued 1994*. Cambridge, MA: MIT Press.

Miller, W. R., & Rollnick, S. (2013). Motivational Interviewing (3rd ed.). New York/London: Guilford Press.

Oldenburg, B., & Parcel, G. (2002). Diffusion of Health Promotion and Health Education Innovations. In K. Glanz, B. K. Rimer, & F. M. Lewis (Eds.), *Health Behavior and Health Education: Theory, Research, and Practice* (3rd ed.). San Francisco: Jossey-Bass.

Otani, Y., Ando, T., Atobe, K., Hauden, A., Kao, S.-Y., Saito, K., & Fukunaga, K. et al. (2012). Comparison of Two Large Earthquakes: The 2008 Sichuan Earthquake and the 2011 East Japan Earthquake. *The Keio Journal of Medicine*, *61*(1), 35–39. doi:10.2302/kjm.61.35 PMID:22410538

Ozaki, R., & Sevastyanova, K. (2011). Going hybrid: An analysis of consumer purchase motivations. *Energy Policy*, *39*(5), 2217–2227. doi:10.1016/j.enpol.2010.04.024

Park, K. W. (2017). *Electric Vehicles: Different Types of EVs and Advantages/Disadvantages*. Charlottesburg, VA: University of Virginia.

Peters, A., Gutscher, H., & Scholz, R. W. (2011). Psychological determinants of fuel consumption of purchased new cars. *Transp. Res. Part F: Psychol. Behav.*, *14*(3), 229–239. doi:10.1016/j.trf.2011.01.003

Plotz, P., Schneider, U., Globisch, J., & Dutsche, E. (2014). Who will buy electric vehicles? Identifying early adopters in Germany. *Transportation Research Part A, Policy and Practice*, *67*, 96–109. doi:10.1016/j.tra.2014.06.006

Prochaska, J. O., DiClemente, C. C., & Norcross, J. C. (1992). In search of how people change: Applications to addictive behaviors. *The American Psychologist*, *47*(9), 1102–1114. doi:10.1037/0003-066X.47.9.1102 PMID:1329589

Prochaska, J.O., Norcross, J.C. & DiClemente, C.C. (1994). *Changing for good*. New York: Morrow.

Prochaska, J. O., Redding, C., & Evers, K. (2009). The Transtheoretical Model and stages of change. In K. Glanz, F. M. Lewis, & B. K. Rimer (Eds.), *Health Behavior and Health Education: Theory, Research and Practice* (4th ed.). Jossey-Bass Publications, Inc.

Rogers, E. M. (1962). *Diffusion of Innovations* (1st ed.). New York: Free Press.

Rogers, E. M. (1983). *Diffusion of Innovations* (3rd ed.). New York: Free Press.

Rogers, E. M. (1995). *Diffusion of Innovations* (4th ed.). New York: Free Press.

Rogers, E. M. (2003). *Diffusion of Innovations* (5th ed.). New York: Free Press.

Simons-Morton, B. G., Greene, W. H., & Gottlieb, N. H. (1995). *Introduction to Health Education and Health Promotion* (2nd ed.). Long Grove, IL: Waveland Press.

Stichler, J.F. (2011). Leading Change. *Nursing for Women's Health, 15*(2), 166-170. doi:10.1111/j.1751-486X.2011.01625.x

Szabo, E. (2014, February 5). CVS decision comes amid renewed anti-smoking efforts. *US Today*. Retrieved from http://www.usatoday.com/story/news/nation/2014/02/05/smoking-cvs-health/5230621/

United Nations. (2015b, November 4). Paris Climate Agreement to enter into force. Retrieved from http://www.un.org/sustainabledevelopment/blog/2016/10/paris-climate-agreement-to-enter-into-force-on-4-november/

United Nations Development Programme. (2015a). Human Development Reports. Table 1: Human Development Index and its components. Retrieved from http://hdr.undp.org/en/composite/HDI

Weiss, D. (2016). Assessing Online Courses in Health Education: Training a 21st Century Health Workforce. In V. Wang (Ed.), Handbook of Advancing Health Education through Technology. IGI Global (Information Service Reference): Hershey, PA. ISBN: 978146669494.1

World Bank Overview of China. (2016). Overview, China. Available at: http://www.worldbank.org/en/country/china/overview

Zhou, H., Shi, L., Yuping, M., Tang, J., & Zeng, Y. (2012). *Diffusion of new technology, health services and information after a crisis: a focus group study of the Sichuan "5.12" Earthquake*. Beijing, China: John Wiley & Sons Ltd.; doi:10.1002/hpm.2137

KEY TERMS AND DEFINITIONS

Adoption: Uptake of a health education program by a target audience.

Bell Curve: A normal distribution that shows the range of values for a given data set (e.g., test scores for a class); approximately 68% of the data lies within one standard deviation (SD) of the mean (average), 95% within two SDs of the mean, and 99.7% within three SDs of the mean.

Cosmopolite: Describes a person who tends to have interpersonal networks outside of, rather than within, his or her social system.

Diffusion: The process by which an innovation is communicated over time among the members of a social system.

Diffusion of Innovations (DOI) Theory: A theory developed by American sociologist Everett M. Rogers that explains how, why, and at what rate innovations are communicated, spread, and adopted by individuals, organizations, and society (see Everett M. Rogers).

Dissemination: Active approach for knowledge transfer from the resource system to the users.

Innovation: An idea, practice, or object that is perceived as new by an individual or other unit of adoption.

Motivational Interviewing: A collaborative conversational style for a strengthening an individual's own motivation and commitment to change.

James O. Prochaska: Dr. Prochaska is one of the originators of the Transtheoretical Model of Behavior Change, a widely replicated model of behavior change (see Transtheoretical Model and Stages of Change); he directs the Cancer Prevention Research Center at the University of Rhode Island.

Everett M. Rogers: (1931-2004): An American sociologist who developed the Diffusion of Innovations Theory, which has been widely used in communications, technology, and health education, including hygiene, family planning, and cancer prevention.

Stages of Change: During the process of making a behavior change, an individual moves through the following five stages: precontemplation, contemplation, preparation, action, and maintenance.

Transtheoretical Model: An integrative, biopsychosocial model that conceptualizes the process of intentional behavior change using five Stages of Change.

Chapter 6
Sustainable Development

ABSTRACT

Sustainable development works within the ecosystems of the planet to conserve and replenish renewable clean energy sources. The most widely used definition of sustainable development was written by Norway's Prime Minister in her 1987 Brundtland Report to the United Nations: it must meet the needs of the present without compromising the ability of future generations to meet their own needs. The United Nations has adopted 17 Sustainable Development Goals, to be achieved by 2030; they include the eradication of poverty and hunger, provision of clean water and sanitation for all, achievement of gender equality, elimination of health disparities, access to quality education, promotion of economic prosperity, use of affordable clean energy sources, and good health and well-being for all. Proper disposal of hazardous waste and effective disaster risk management help to facilitate these goals. This chapter will discuss several new technologies employed in sustainable development.

INTRODUCTION

In 1987, the Chair of the World Commission on Environment and Development, Gro Harlem Brundtland, the Prime Minister of Norway, presented the commission's report to the United Nations. This landmark report was the result of four years of work by the World Commission. In 1983, the Secretary-General of the United Nations had charged the World Commission to accomplish the following: 1) propose long-term environmental strategies for achieving sustainable development; 2) foster greater co-operation on

DOI: 10.4018/978-1-5225-3414-3.ch006

environmental concerns among developing countries and between countries at different stages of economic and social development; 3) consider how the international community could deal more effectively with environment concerns; and 4) define shared perceptions of long-term environmental issues and the appropriate efforts needed to deal successfully with the problems of protecting and enhancing the environment (Brundtland, 1987).

The Brundtland report set forth the following definition of sustainable development:

- Development that meets the needs of the present without compromising the ability of future generations to meet their own needs (Brundtland, 1987, Chairman's Foreward).

It contains within it two key concepts:

- The concept of needs, in particular the essential needs of the world's poor, to which overriding priority should be given;
- The idea of limitations imposed by the state of technology and social organization on the environment's ability to meet present and future needs (Brundtland, 1987, Chairman's Foreward).

In Part IV of the Brundtland Report, *A Call to Action*, the World Commission compared humans' journeying into outer space in the 20th century to the Copernican revolution, which changed human thinking about the place of our planet in the universe; people understood that the earth circled the sun, not the reverse. Viewing earth from outer space, human activity is not visible; rather, what is seen is a pattern of clouds, oceans, greenery, and soils on a small fragile ball. The Brundtland Report (1987, p. 11) warned of the dangers created by human disregard of the earth's ecosystems:

Humanity's inability to fit its activities into that pattern is changing planetary systems, fundamentally. Many such changes are accompanied by life-threatening hazards. This new reality, from which there is no escape, must be recognized - and managed.

On September 15, 2015, the United Nations countries adopted the following 17 sustainable development goals to end poverty, protect the earth, and ensure prosperity for all people, with specific targets to be achieved by 2030:

1. No poverty.
2. Zero hunger.
3. Good health and well-being.
4. Quality education.
5. Gender equality.
6. Clean water and sanitation.
7. Affordable and clean energy.
8. Decent work and economic growth.
9. Industry, innovation, and infrastructure.
10. Reduced inequalities.
11. Sustainable cities and communities.
12. Sustainable consumption and production.
13. Climate action.
14. Sustainable life below water.
15. Sustainable life on land.
16. Peace, justice, and strong institutions.
17. Partnerships for the goals (United Nations, 2017).

The first step in fostering sustainable development is to have a clear understanding of the principles of sustainability.

PRINCIPLES OF SUSTAINABILITY (REDUCE, REUSE, RECYCLE)

Three basic principles –reduce, reuse, recycle-- form the underpinning of sustainable development. Just as the three R's of education -- reading, writing, and arithmetic -- provide building blocks for a person's lifelong learning, the three R's of sustainability provide a foundation for the world's current and future socioeconomic development. Renewable, reusable resources go back into the mix, to be re-created and used again. In countries where resources such as clean water are scarce, residents have learned to reduce usage and conserve energy. Nearly 1 billion people in the developing world today don't have access to clean drinking water. Even though 70 percent of the Earth's surface is covered with water, 97.5 percent of that is saltwater. Of the 2.5 percent that is freshwater, 68.7 percent is frozen in ice caps and glaciers (The Water Project, 2017).

In the United States, which has long been a country with abundant water and natural resources, much more can be done to reduce natural resource consumption and to reuse and recycle to create less waste. The landfills are running out of space, piled high with mounds of Styrofoam, Pampers, plastic bags, food waste, and other debris. The most effective way to deal with excess waste is to not create it in the first place. Creating a new product requires a lot of energy and raw materials, which must be extracted from the earth; then the product must be manufactured and transported to wherever it will be sold. Reduction and reuse are the most effective ways people can save natural resources, protect the environment, and save money.

Following are recommendations for water conservation at home:

- Install energy-efficient low-flow toilets to reduce water usage by 40 percent to 50 percent.
- Install low-flow shower heads or adjustable flow-reducer devices on shower head to reduce water flow by 35 percent.
- Install an aerator and/or a water flow-reducer attachment on the tap to reduce water usage.
- Turn off tap tightly to avoid drips.
- Check pipes, sinks, and tubs for leaks and promptly repair any leaks.
- Run washing machines with full loads on shortest cycle using energy-save.
- Run dishwashers on shortest energy-saver cycles with full loads.
- When hand-washing dishes, never run water continuously; wash dishes in a partially filled sink and then rinse using the spray attachment.
- When brushing teeth, turn the water off while brushing and use short bursts of water to clean brush.
- When washing or shaving over a sink, partially fill the sink and use that water rather than running water continuously.
- Use eco-friendly cleaning products.
- Water lawns every three to five days during the cool part of the day and use timers to shut off sprinklers; allow 5 millimeters of water for each day since the last watering (The Water Project, 2017).

To reduce waste, buy reusable over disposable items. Bring reusable bags to do grocery shopping, and fill up reusable water bottles and coffee cups rather than using disposable plastic or styrofoam. Bring silverware to work rather than using plastic utensils. Look for products that use less packaging, as they use less raw material. Buy in bulk. Maintain and repair clothing, tires and appliances, so that they won't have to be replaced as frequently; borrow items that are used infrequently. Donate used items to charities rather than discarding them. Alternatively, trade or sell them on line.

Recycling is the third leg of sustainable development. In 1980, recycling and composting saved 14.5 million tons of trash from going into landfills and incinerators in the United States. By 2013, 87.2 million tons of trash was being recycled, six times as much as in 1980. However, that figure still represents only 34.2 percent of the nation's garbage (Denchak, 2016). Recycling starts with consumer purchasing: buy products that are recyclable and are made of recycled goods. Recycled products are easily identified by the universal symbol for recycling (see Figure 1).

Know your local regulations about recycling, as they vary from community to community. Follow the guidelines for recyclables. For example, rinse out bottles and cans before recycling. Do NOT recycle plastic containers in plastic bags, as the bags will clog up the machines at the recycling centers. Plastic bags may be recycled in separate bins at some venues such as supermarkets. To minimize plastic bag use, bring your reusable bags with you to shop for food, clothing, and household items.

Today, electronics such as personal computers (PCs), printers, and mobile devices can be recycled at major retailers. Computer repair stores will often buy back and salvage old computers for parts. Table 1 list some major retailers that recycle mobile phones, PCs, and televisions.

Figure 1. Universal symbol for recycling

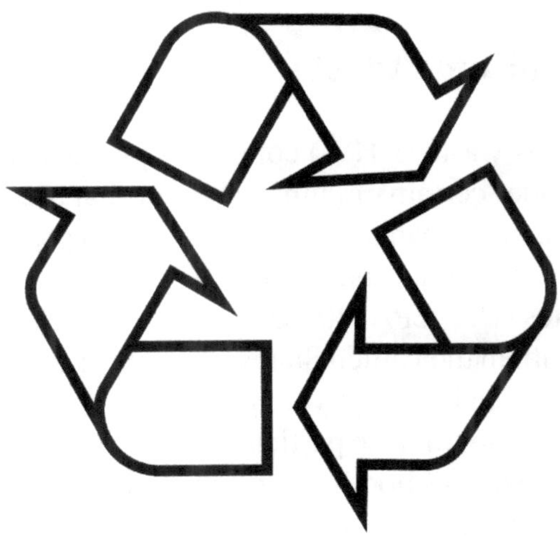

Table 1. Recycling options for mobile devices, PCs and TVs

Electronic Device			Store	What Store Offers
Mobile device	Personal Computer	Television	Best Buy	In-store, event, and online recycling options, plus haulaway for TV
Mobile device	Personal Computer	x	Dell	In-store and event recycling, plus permanent drop-off site and mail-in for PC
Mobile device	x	Television	LG	Mail-in recycling for mobile, event recycling and permanent drop-off site for TV
Mobile device	Personal Computer	Television	Samsung	Mail-in recycling for mobile, permanent drop-off site recycling options for PC, TV
x	Personal Computer	Television	Sony	In-store, permanent drop-off site, and event recycling options for PC, plus haulaway for TV
Mobile device	x	x	Sprint	In-store, permanent drop-off site and mail-in recycling options
Mobile device	Personal Computer	Television	Staples	In-store and event recycling options
x	Personal Computer	Television	Vizio	Mail-in recycling options and permanent drop-off site in some states

Source: U.S. Environmental Protection Agency (2017a)

DISPOSAL OF HAZARDOUS WASTE

Hazardous waste must be disposed of properly, in adherence with safety regulations, to prevent injury and the release of toxic chemicals into the atmosphere. Individuals are responsible for the safe disposal of household hazardous waste; businesses are responsible for the disposal of their industrial hazardous waste. The following section provides guidelines for reducing usage of household hazardous waste and disposing of it safely.

Household Hazardous Waste

Household hazardous waste (HHW) comprises products that can catch fire, react, or explode under certain circumstances, or that are corrosive or toxic. Examples include paints, cleaners, oils, batteries, and pesticides, all of which contain hazardous ingredients and must be disposed of at special hazardous waste collection sites (U.S. EPA, 2016a). Poisoning is the number one cause of unintentional death in the United States; poison control centers receive 2.2 million calls a year. Toddlers and young children explore by touching and putting things in their mouth. One particularly dangerous item that appeals to young children because it is smooth and a shiny silver color is button batteries, which are used in hearing aids, watches, toys, games, flashing jewelry, singing greeting cards, and remote control devices. If a button battery is swallowed or placed in the nose or ears, it can cause serious injury or death. In the U.S., more than 3,500 people a year swallow button batteries (National Capital Poison Center, 2017). They can get stuck in the esophagus and cause tissue burns that can be fatal.

The National Capital Poison Center in Washington, D.C. operates a 24/7 hotline for battery ingestion cases. Call 202-625-3333 immediately if someone you know has swallowed a battery. If possible, provide the battery identification number, found on the packaging. The person who swallowed the battery should not eat or drink or induce vomiting until after an X-ray has been taken, to make sure that the battery has passed through the esophagus to the stomach. If the battery remains in the esophagus, it must be removed immediately (National Capital Poison Center, 2017). Report any of the following symptoms—fever, abdominal pain, vomiting, or blood in the stool. Watch for the battery to pass in the stool. Button batteries can also cause permanent injury if lodged in the nose or ear. Symptoms to watch for are pain or discharge from the nose or ear. Do NOT use nose or eye drops until the person has been examined by a healthcare provider, as drops may cause additional injury if they contact a battery (National Capital Poison Center, 2017).

Dispose of hazardous waste products safely by bringing them to a household hazardous waste collection site near you. To find a site in your zip code, search in the Earth 911 database (http://search.earth911.com/) for information. Improper disposal of HHW products, such as pouring them down the drain, on the ground, into storm sewers, or putting them out with the regular trash, is hazardous to human health and to the environment. Such improper disposal may cause physical injury to sanitation workers, and may contaminate septic tanks or wastewater treatment systems if poured down drains or toilets (U.S. EPA, 2016a). Leftover HHW products, such as cans of paint and pesticide, may pose dangers to children and pets if left around the house.

To minimize risk of health hazards from HHW products, follow these safety tips:

- Carefully read and follow product label instructions for use, storage, and disposal.
- Do NOT store hazardous products in food containers; keep them in their original containers and never remove labels. If containers become corroded, call your local hazardous materials official or fire department for special handling instructions.

- DO NOT mix leftover HHW products with other products, as they might react, ignite, or explode, and become contaminated, which renders the HHW unrecyclable.
- Check with your local environmental, health or solid waste agency for HHW management options in your area.
  - If your community doesn't have a year-round HHW collection system, check if there are any designated days for collecting HHW at a central location.
  - If your community has neither a permanent collection site nor a special collection day, check if recycling or disposal is available at local businesses; for example, some local garages may accept used motor oil for recycling.
  - Handle empty containers of HHW with care, as they can pose hazards because of residual chemicals (U.S. EPA, 2016a).

The fewer HHWs in your household, the easier to manage their use, storage, and disposal. Consumers can reduce the number of HHW products in their homes by using alternative products; the U.S. Environmental Protection Agency has a list of approved "Safer Choices" products (https://www.epa.gov/saferchoice). The number of HHW products also can be reduced following the recommendations in Table 2.

Improper disposal of industrial hazardous waste can lead to pollution of soil, water, and air, and may result in cancer and other illnesses in those exposed to it. The following section describes United States guidelines for the

Table 2. Reducing usage of household hazardous waste products

HHW to Avoid	Alternative to Use Instead
Drain Cleaner	Use a plunger or plumber's snake
Glass Cleaner	Mix one tablespoon of vinegar or lemon juice in one quart of water, spray on, and use newspaper to dry
Furniture Polish	Mix one teaspoon of lemon juice in one pint of mineral or vegetable oil and wipe furniture.
Rug Deodorizer	Liberally sprinkle carpets with baking soda, wait at least 15 minutes and vacuum, repeating if necessary
Silver Polish	Boil 2-3 three inches of water in a shallow pan with one teaspoon of salt, one teaspoon of baking soda and a sheet of aluminum foil, submerge silver and boil for two to three more minutes; wipe away tarnish and repeat if necessary.
Mothballs	Use cedar chips, lavender flowers, rosemary, mints or white peppercorns.

Source: U.S. Environmental Protection Agency (2016a)

safe and proper disposal of industrial hazardous waste. To be in compliance, hazardous waste managers must follow specific regulations governing the disposal of industrial hazardous waste.

Industrial Hazardous Waste

Recycling hazardous waste conserves resources and protects the environment. It is cost-effective and discourages toxic dumping, as was the case with the DDT contamination of water in Triana, Alabama, by the Olin Corporation (Case Study #1, Chapter 4), and with the PCB contamination of the soil in Chicago (Case Study #2, Chapter 4). When industrial waste is recycled, its regulatory classification (i.e., whether or not it is a solid waste, and potentially a regulated hazardous waste) depends on what type of secondary material is being recycled and what type of recycling is being done. The U.S. Environmental Protection Agency has tailored the level of regulation to reflect the hazard of the recycling activity. It is up to the waste generator to determine whether his or her recyclable secondary material is subject to reduced requirements or full Resource Conservation and Recovery Act (RCRA) Subtitle C regulation by checking the U.S. Environmental Protection Agency's Hazardous Waste Recycling Regulations (U.S. EPA, 2016b). The RCRA Subtitle C established a federal program to manage hazardous waste generation, transport, treatment, storage, and disposal to ensure protection of human health and the environment. The RCRA Subtitle C program conducts investigations, works with State agencies and the Department of Justice to enforce hazardous waste laws and apprehend those who violate them, and helps state and local agencies administer hazardous waste management programs (U.S. EPA, 2016b).

With some qualifications, the following materials, when recycled, are not considered solid waste and therefore are not subject to RCRA Subtitle C regulation:

- Pulping liquors (black liquor),
- Spent Sulfuric Acid,
- Closed-Loop Recycling,
- Spent Wood Preservatives,
- Coke By-Product Wastes,
- Splash Condenser Dross Residue Hazardous Oil-Bearing Secondary Materials and Recovered Oil from Petroleum Refining Operations,
- Processed Scrap Metal,

- Shredded Circuit Boards,
- Condensates from Kraft Mill Steam Strippers,
- Comparable Fuels,
- Mineral Processing Spent Materials,
- Petrochemical Recovered Oil,
- Spent Caustic Solutions from Petroleum Refining Hazardous Secondary Materials Used to Make Zinc Fertilizers and Zinc Fertilizers Made from Recycled Hazardous Secondary Materials,
- Used Cathode Ray Tubes (CRTs),
- Hazardous Secondary Materials Generated and Reclaimed Under the Control of the Generator,
- Hazardous Secondary Materials Transferred Off-site for Reclamation (U.S. EPA, 2016b).

The following materials, when recycled, are excluded from the definition of hazardous waste and are not subject to hazardous waste regulation:

- Agricultural Waste,
- Spent Chlorofluorocarbon Refrigerants,
- Used Oil Filters,
- Used Oil Distillation Bottoms (U.S. EPA, 2016b).

With some limitations, the following hazardous waste materials are not subject to RCRA Subtitle C regulation when recycled:

- Industrial ethyl alcohol,
- Scrap metal,
- Waste-derived fuels from refining processes,
- Unrefined waste-derived fuels and oils from petroleum refineries (U.S. EPA, 2016b).

Generators should read the regulations carefully to ensure the safe and proper recycling or disposal of hazardous waste.

DISASTER RISK ANALYSIS

In his opening remarks at the launch of the report for the United Nations World Humanitarian Summit in 2016, Secretary General Ban Ki-Moon stated that

We must move from managing crises to preventing them... We must also anticipate crises. That means investing in data and risk analysis, acting early on this information and managing risk before and after crises... to reduce people's vulnerability and risk (Bi-Moon, 2016).

New technologies facilitate data and risk analysis, and enable a more effective disaster response. Risk analysis considers both the probability of a disaster occurring, and the exposure of the elements to it; disaster risk assessment examines the factors that cause losses in order to estimate loss probabilities (Dilley et al., 2005). Global disaster databases help emergency managers to monitor where disasters tend to strike, and to target high-risk regions in emergency preparedness efforts. These databases are getting more sophisticated and portable, allowing researchers in the field to access them on handheld or mobile devices. Another useful technology which has found an innovative new role in disaster management is drones, automated unmanned aircraft that can map damage during and after a disaster, and aid in search-and-rescue operations.

Global Disaster Databases

The United Nations Office for Disaster Risk Reduction (UNISDR) supports the global development of national disaster loss databases; the collection, analysis, and management of such data can help define both short and long-term development goals, and help to identify and address disaster risks. New technologies have allowed public and private sector organizations to capture, store, and analyze data in a structured, systematic manner. Three technologies are of particular importance:

- Global Assessment Report,
- DesInventar,
- EM-DAT: The International Disaster Database.

The Global Assessment Report focuses on disaster risk reduction and the promotion of sustainable development. Its 2015 publication, the 4th edition, clearly linked UNISDR's twin goals, stating, "Sustainable development cannot be achieved unless disaster risk is reduced" (Global Assessment Report, 2015, p. v). The 4th edition warned of growing risk inequality and noted that disaster management was best approached as a three-pronged risk management: 1) prospective risk management, which aims to avoid the accumulation of new risks; 2) corrective risk management, which seeks to reduce existing risks; and 3) compensatory risk management, which support the resilience of individuals and societies as they face residual risk that cannot be effectively reduced. As for making development sustainable, the authors noted (Global Assessment Report, 2015, p. xvi):

If an accelerated increase in disaster risk is to be avoided, there is a growing consensus that the development drivers of risk, such as climate change, the overconsumption of natural capital, poverty and inequality will have to be addressed.

The 4th edition, known as GAR15, included the following technological enhancements, which can be accessed on a Smartphone, tablet or PC:

- **The Earth Icon:** Links user to a dynamic 3D globe, enabling geospatial data;
- **The Researcher Icon:** Links user to data about the researchers behind the article;
- **The Tablet Icon:** Links user to dynamic animations and other information on print charts;
- **The Video Icon:** Links user to relevant videos of UNISDR and partners;
- **The Web Link Icon:** Links users to external websites [Global Assessment Report, 2015].

DesInventar is a Disaster Information Management System that generates National Disaster Inventories and enables the systematic analysis of disaster trends and their impacts. EM-DAT is an international database that provides core data on natural and technological disasters from 1900 to the present, compiled from multiple sources (UNISDR, 2017).

Registries and COINs

One of the keys to successful emergency response is maintaining an up-to-date complete database of names, addresses, and contact information of everyone who lives in the community, and to include in this database key information on age, physical limitations or special needs of the residents. Registries are useful tools for identifying and reaching all members of a community. To maximize cooperation, emergency managers should conduct their registry outreach through trusted community organizations. Information may be collected by volunteers or social service workers when individuals apply for other public health services. Registry information also may be collected by phone, on line, by having a form mailed with a postage-paid addressed envelope for easier return, or door to door. It may seem old-fashioned in our digital world, but door-to-door canvassing is still a respected and oft-used method of surveying a neighborhood. Social media, TV, newspapers, and brochures all may be used to advertise a registry and encourage people to sign up for it. Signing up for a registry can be piggybacked on other community events such as a local health fair or festival. Registries need to be updated annually, as people move, die, or their vulnerability status changes.

A Community Outreach Information Network (COIN) is a grassroots network of trusted community leaders who can help locate and sign up at-risk individuals for an emergency database. The COIN will be helpful in notating the languages spoken at home, cultural practices, beliefs, and physical or mental limitations of community members. In addition, local nonprofits and faith-based organizations can help reach out to isolated at-risk individuals (CDC, 2015).

Drones

Drones are unmanned aerial vehicles (UAVs) that have found a new use in disaster management. Originally used for military operations, drones are now used in monitoring disasters, mapping areas that have been struck by disasters, and in search-and-rescue missions to aid victims of disasters. Drone mapping also is utilized in disaster risk reduction, as the following case example illustrates.

Case Example 1: Flood Mapping for Disaster Risk Reduction in Tanzania

More than 70 percent of Dar es Salaam's 5 million people live in informal, unplanned settlements with inadequate infrastructure (Swiss Foundation for Mine Action, 2015). Twice a year heavy rainfalls result in significant flood risks. The objectives of drone mapping in Dar es Salaam were to use flood mapping to reduce disaster risk by obtaining high-quality exposure maps of affected communities and then to create a hydrological model using elevation data to manage flood risks. New innovations in digital photography that allowed for high-quality digital imagery contributed to the success of the flood mapping project (Swiss Foundation for Mine Action, 2015).

In 2015, with the support of the World Bank and Red Cross, the Ramani Huria (Open Mapping) coalition was formed by the City Council of Dar es Salaam, the Tanzania Commission for Science and Technology (COSTECH), the University of Dar es Salaam, Ardhi University and Buni Innovation Hub. The coalition's goal was to map flood-prone areas in Dar es Salaam, to make the mapping data openly available, and to promote the use of this information in government decisions to mitigate flooding (Swiss Foundation for Mine Action, 2015). The World Bank invited Drone Adventures, a non-profit organization based in Switzerland, to test the idea of collecting and using drone imagery for this mapping project. The mapping was done in two phases and covered an area of 88 square km; one pilot operated 2-3 eBee drones, which flew simultaneously over the mapping area. Drone Adventures used eBee Image processing software, Postflight Terra 3D to patch the separate images together to create two main datasets--an optical 2D orthomosaic model and a 3D surface model--which were processed in Switzerland. The next step was to convert the aerial imagery into digital street maps. Finally, using software called LAStools, the buildings and other artificial structures were removed to create a model of the underlying terrain; the team is still working on the hydrological model. The project highlighted infrastructure vulnerabilities, which the government is addressing. There was concern about the social acceptance of using drones for mapping over an urban area. Interviews were conducted with 14 high-level government officials and 208 community members who witnessed drone flights over the city, and both groups reported positive feedback about the use of drones for humanitarian purposes such as protecting residents from flooding (Swiss Foundation for Mine Action, 2015).

Case Example 2: Using Drones to Inspect Post-Earthquake Road Damage in Ecuador

In April 2016, a magnitude-7.8 earthquake struck north-west Ecuador, killing more than 660 people and damaging homes, buildings, and infrastructure such as roadways and bridges (Telesur, 2016). A state of emergency was declared in the six coastal provinces. In the two days following the earthquake, there were 189 aftershocks and the soil shifted; thousands were left homeless (DuPlessis, 2016). Volunteer drone operators offered to help the government inspect the damage and make a risk assessment. The drones included two quadcopters – the DJI Phantom 3 Professional and DJI Inspire 1 – and several fixed-wing eBees. The Phantom 3 can fly for 23 minutes and up to 5 km away from its pilot. The Inspire can fly for 15 minutes and up to 5 km from its pilot. The eBee, a widely used mapping drone, is a fixed-wing craft and able to cover 12 square km in one flight, with a flying time of about 50 minutes (DuPlessis, 2016). Within days, the 10-12 drones utilized in mapping the earthquake damage had collected data and published it on a public geo-portal platform run by OpenAerialMap (affiliated with OpenStreetMap). Their swift response provided decision makers with critical tools in the time period immediately following the disaster. Geologists assisted drone operators in assessing damage to roadways, and to the surrounding hillsides and slopes, in order to prioritize the response. The drones were able to map remote and hard to reach areas not accessible by roadways. The aerial maps provided proof of damage and justification for investment in emergency response (DuPlessis, 2016).

The use of innovative technology—WhatsApp and Google Drive--allowed the drone pilots to cover a large area every day without duplication. The pilots coordinated deployments and discussed launch times and locations using WhatsApp. For each mission, deployment locations and times were updated and shared on Google Drive (DuPlessis, 2016).

The rapid response of the drones after the earthquake was possible largely because drones were already on the ground prior to the earthquake, and their pilots had been trained on the use of drones in humanitarian response operations. The drone as a mapping tool was easily integrated into the general assessment toolkit at the Ministry of Transport (DuPlessis, 2016).

WATER MANAGEMENT

In 2012, the National Intelligence Council of the United States issued a report on Global Trends that concluded that the world is entering into a prolonged water crisis. The report predicted that countries important to the US and global security would be at risk of "state failure" in less than a decade. Twenty percent of the world's population—about 1.5 billion people--will suffer from this water crisis, and 600 million have already begun to experience water shortages. Water shortages are expected to impede energy production, threatening global food markets and slowing economic growth (Office of the Director of National Intelligence, 2012). World powers India and China, as well as California, are already experiencing water shortages and droughts. Parts of the San Joaquin Valley, which produces more grapes, oranges, peaches, vegetables, almonds, and pistachio nuts than anywhere else in the US, are running out of water. In addition, water safety is at issue from pesticide use (see Chapter 4, Case Example 1, DDT contamination in Triana, Alabama). Other regions of the US are also experiencing water shortages. The High Plains Aquifer, which provides water to farms in Colorado, Nebraska, Kansas, and Texas, has been depleted by overpumping (Zwingle, 1993), and the water in Lake Mead is getting dangerously low, which could affect the production of clean hydroelectric power for Southwestern States (Wines, 2014). Pollution is also limiting clean water supplies; Florida's largest source of freshwater, the Manatee Springs and Aquifer, has been polluted by agricultural runoff and needs treatment (Alvarez, 2014).

The following five macro trends are driving this worldwide water crisis (Siegel, 2015):

- **Population:** The world's population is 7.5 billion (Worldometers, 2017) and is projected to increase by more than one billion people within the next 15 years, reaching 8.5 billion in 2030, 9.7 billion in 2050, and 11.2 billion by 2100 (United Nations, 2015).
- **Rising Middle Class:** Globally, the middle class is expanding; there were 1.4 billion middle class people in 2000, more than 1.8 billion in 2009, and by 2020, there is projected to be 3.25 billion middle class (Kharas & Gertz, 2010). When people move up from poverty to middle class, they consume much more water. Diets of those living in poverty tend to be vegetable- and grain-based, whereas diets of middle-class people are protein-based. Raising a pound of beef uses 17 times more water than growing a pound of corn (Office of the Director of National Intelligence, 2012).
- **Climate Change:** Rising temperatures of water surfaces due to climate change accelerate evaporation of water (Gronewold & Stow, 2014). Precipitation patterns have been changing, with less frequent and more intense rainfalls, which has led to the hardening of soil and reduced penetration of water into the ground.
- **Tainted Water:** Globally, contaminated water and poor sanitation are linked to transmission of diseases such as cholera, diarrhoea, dysentery, hepatitis A, typhoid and polio; approximately 842,000 deaths per year are due to unsafe drinking-water, sanitation and hand hygiene (WHO, 2016). Almost 240 million people have schistosomiasis, which is contracted from parasitic worms in infested water. Inadequate

management of urban, industrial and agricultural wastewater means the drinking-water of hundreds of millions of people is dangerously contaminated or chemically polluted (WHO, 2016).

- **Leaks:** An immense amount of water is lost every day in municipalities due to leaks, open fire hydrants, theft, and neglect (Siegel, 2015). Every year, London loses about 30 percent of its water, and Chicago, about 25 percent of its water (Smart Water Networks Forum Research, 2010). In major middle Eastern and Asian cities, up to 60 percent of the water is lost yearly due to faulty infrastructure (Al-Ansari et al., 2014).

The Water Revolution in Israel

Israel's water revolution has turned around a drought-stricken land that is 60 percent desert into a world leader in water management. Its five water management principles are 1) education and conservation, 2) wastewater reclamation and reuse (86 percent of wastewater reused for agriculture), 3) government policy (centralized water control), 4) water and agriculture technology (e.g., drip irrigation), and 5) desalination (provides 25 percent of nation's water supply, 80 percent of household water) (Distel, 2015).

Education and Conservation

From its infancy, Israelis have practiced water conservation; it is taught to schoolchildren from an early age. Every activity involving water—from brushing teeth to showering—is done in an energy-efficient manner to use as little water as possible. When the price of water was raised in 2008, Israelis became even more conservative and efficient with their water use (Siegel, 2015).

Wastewater Reclamation and Reuse

Israel is the world leader in recycling and reusing wastewater for agriculture; it treats and reuses an astonishing 86 percent of its domestic wastewater for agricultural use; plans are to increase this to 95 percent by 2025. In Israel, 60 percent of fruits and vegetables are grown using recycled, purified wastewater (Distal, 2015). This frees up a significant amount of freshwater for drinking and other purposes. Reducing the amount of fresh water needed for agriculture is vitally important, as agriculture accounts for by far the greatest portion of global water usage--about 68 percent of total water usage, compared to about 20 percent for households and individuals, and about 10 percent for power stations (Hazony, 2015).

Government Policy: Centralized Water Control

In Seth Siegel's 2015 book, *Let There Be Water*, he describes the Israeli decision to make water the common property of all and to regulate it through one centralized water authority. In the mid-1950s, three important water laws were passed by the Knesset, Israel's Parliament: the first, in 1955, prohibited any drilling for water in Israel, even by an owner on private property, without first obtaining a license; the second law, also passed in 1955, prohibited distribution of water without a meter and that all utilities install separate meters to measure the amount of water provided to each home or business. In 1957, the third water law placed all surface water--in lakes, rivers, and even rainfall—under government control. Permits were required to divert any surface water sources (Siegel, 2015). The centralization of water control was institutionalized in 1959, with the passage of the Water Law of 1959, which vested in the government power to control and restrict activities of individual water users to further and protect the public interest. The Israeli government built infrastructure to bring water from the north to the arid south—an expensive and risky undertaking that succeeded and made the Negev desert bloom (Siegel, 2015).

Water and Agricultural Technology

Unlike most utilities, the Israel Water Authority has always turned to technological innovation to boost efficiency and productivity. One important cost-saving innovation was the development and implementation of Distant Meter Reading (DMR) technology. DMR checks water usage every four hours. It eliminates the need to pay meter readers to come to each home or business. It develops a "consumption fingerprint" for each household or business so that consumption outside of normal usage is a red flag. This innovative technology catches leaks early on, saving a great deal of water and money (Siegel, 2015).

The greatest technological innovation in agricultural water management that Israel has given the world is drip irrigation, which was invented by Israeli water pioneer Simcha Blass, who noticed that one tree in a line of trees was taller and more robust than the others. He investigated and found that there water was dripping onto the tree from a leaky pipe. In contrast to flood irrigation, which floods irrigation ditches with water, drip irrigation is more efficient, supplying the water in a steady drip (Siegel, 2015). Drip irrigation not only conserves water, it also produces more crops, as the plants can absorb the water better than in flood irrigation. It took years to perfect and implement drip irrigation in Israel; Blass patented drip agriculture in 1959 and partnered with Kibbutz Hatzerim in 1964 to create the irrigation company Netafim (Siegel, 2015). Today drip irrigation is widely considered one of the great agricultural innovations of the 20th century, and it has been adopted in many countries around the world (Siegel, 2015). Israeli donations of drip irrigation kits, seed, and fertilizer to farmers in developing countries in sub-Saharan Africa and Asia help to fight hunger and alleviate poverty (Nourishing the Planet, 2010).

Drip agriculture technology also provides a remedy for the toxic algae blooms that have been proliferating as a result of warmer temperatures combined with agricultural fertilizer runoff in freshwater lakes and rivers. The nitrogen and phosphorus in the fertilizer provide a food source for algae blooms, which explode after a few days on warm temperatures. Toxic blue-green algae blooms seen on lakes cause foul odors, contaminate the water so that it is not potable, and suck the oxygen out of the water, killing off plants and fish. Drip irrigation technology can be employed to water and fertilize plants at the same time using water-soluble fertilizer. This process is called fertigation, and it uses much less fertilizer than do traditional techniques, reducing the risk of runoff and algae blooms in freshwater sources (Siegel, 2015).

Desalination (Desalinization)

In drought-stricken countries such as Israel, where there is a historic lack of sufficient freshwater, desalination becomes a viable option for providing an additional source of freshwater. However, taking the salt out of seawater is costly. The Israel Desalination Engineering team invented two new methods--Mechanical Vapor Compression (MVC) and Multi-Effect Distillation (MED)--for desalination (Siegel, 2015). MED had been used previously to evaporate fruit juice to obtain its natural sugar. The innovation in the Israeli MED approach was to replace the chambers used to heat vapor with linked aluminum tubes, which transferred heat more effectively than any previous material that had been used. The aluminum tubes kept the temperature high enough that no additional energy source was needed to heat water during the process, making the Israeli MED process energy-efficient (Siegel, 2015). Funding was finally secured, with the help of the United States, and a demonstration plant was built in Ashdod in the mid-1980s. Another method of desalination—reverse osmosis—was invented by an American chemical engineer who moved to Israel. It took many years to convince a plant to adopt this innovation. The reverse-osmosis method was finally implemented at a plant in Ashkelon in 2005, and proved to be both cheaper and more effective than previous methods, creating freshwater that had higher clarity. Three more reverse-osmosis desalination plants were opened along the Israeli Mediterranean coast in 2007, 2009, and 2013; reverse-osmosis is the dominant method today, used by 60 percent of desalination plants worldwide . The Israel Desalination Engineering, now called IDE Technologies, has built desalination plants in China, India, and around the world (Siegel, 2015). Today, Israel's five desalination plants provide about 25 percent of Israel's water needs and about 50 percent of its household water. Desalination is one element of Israel's innovative, technologically sophisticated, integrated water management system (Distel, 2015).

There is a caveat here—desalination could be harmful to ecosystems, especially from dumping of the very salty sludge left over after salt is removed from the seawater. Just as disposal of hazardous waste must be regulated, so should the waste from desalination be regulated and disposed of so as to protect the environment (Scientific American, 2017). Israel regulates all aspects of water management—so must all countries using desalination technologies.

LAND MANAGEMENT

Land management is another area seriously impacted by global warming. Sea level rise in coastal areas has led to erosion of beaches and flooding of streets during biannual King Tides and storm surges. Droughts and desertification have reduced agricultural yield, causing malnourishment and displacement of environmental refugees. Rising temperatures from global warming, combined with droughts, have led to more frequent and extensive wildfires, which travel at speeds of up to 14 miles an hour (23 kilometers an hour), consuming everything in their path (National Geographic, 2017).

Deforestation in Brazil has eliminated almost 20 percent of the Amazon rainforest in the past 40 years. If this pace continues, the rainforest's ecosystems will begin to fail (Wallace, 2017). A healthy rainforest system produces half of its own rainfall through moisture released into the atmosphere. Deforestation greatly reduces the amount of moisture released into the atmosphere; without enough rainfall, the remaining trees will dry out and die, providing fuel for wildfires (Wallace, 2017). Brazil is the world's eighth largest emitter of greenhouse gases (GHG), and the third largest emitter in the developing world after China and India. Most of Brazil's GHG emissions come from deforestation, both from the burning of the forests and from the clearing of the land, which releases CO_2 and methane held in the soil (La Rovere & Pereira, 2014).

In 2014, the IPCC reported that Agriculture, Forestry, and Other Land Use (AFOLU) accounted for 24 percent of global anthropogenic (i.e., due to human activities) GHG emissions. AFOLU's anthropogenic GHG emissions were primarily from deforestation, livestock, and soil and nutrient management (Smith et al., 2014). Two agricultural innovations--drip agriculture and fertigation--should have more widespread adoption worldwide, as they conserve water, improve crop yield, and reduce pollution from fertilizer runoff, which can cause explosions of toxic algae blooms. Reducing the number of trees cleared can be achieved by legally designating protected areas and reserves, and through the passage of national initiatives to combat illegal logging. Agricultural emissions can be reduced by the widespread adoption of "no-till farming." Instead of turning over the soil with tractors and plows, "no till" farming leaves the stalks of plants in the ground, sequestering the carbon underground rather than releasing it (Smith et al., 2007).

Changing food consumption patterns by reducing reliance on meat is an innovative—and economical—way to reduce GHG emissions and land degradation. .

Reducing Meat Consumption

A cost-effective innovation that has been slowly gaining favor is modifying one's diet to reduce meat consumption, and thereby to reduce energy consumption and GHG emissions. In the average United States diet, animal products make up 60 percent of emissions. The Food and Agriculture Organization of the United Nations (FAO, 2006) estimated that livestock accounted for 18 percent of global GHG emissions, more than the transport sector. Livestock also are a major source of land and water degradation; deforestation has cleared away 70 percent of former forests in the Amazon to convert the land into pastures, and 20 percent of the pastures are degraded through overgrazing, compaction and erosion (FAO, 2006). In addition, methane emitted by animal waste and by the animals themselves as they digest their food accounts (Fiala, 2009) for 37 percent of all human-induced methane, and 64 percent of ammonia, which contributes significantly to acid rain (FAO, 2006). Figure 2 compares the CO_2 equivalent per pound of food produced for meat and other foods.

A half-pound of beef produces 7.4 pounds of CO_2 equivalent emissions, almost 4 times that of pork, more than 13 times that of chicken, 27 times that of asparagus, 37 times that of apples, and nearly 44 times that of potatoes (Fiala, 2009)! Even a minor change in diet—say, substituting chicken for beef half the time and going "meatless" one day a week—would make a significant difference in a person's carbon footprint. Eating primarily a vegetable and grain-based diet would be even better. Of course, vegans leave the smallest carbon footprint, as they use no animal products at all.

The Meatless Monday Movement is a global food initiative with a simple message: once a week, cut out the meat. It was first introduced during World War I as a form of food rationing, and was again implemented during World War II. It was re-introduced in 2003 as an environmentally friendly health prevention initiative by the Center for a Livable Future at the Bloomberg School of Public Health at Johns Hopkins University in Baltimore (Meatless Monday, 2017). It has a twofold aim—to reduce individual carbon footprints and to reduce an individual's risk factors for cardiovascular disease, stroke, cancer, and obesity. Meatless Monday's goal is to reduce meat consumption

Figure 2. Comparison of CO_2 Equivalent Emissions per 1/2 Pound Food Produced
Figure adapted from data in Fiala (2009).

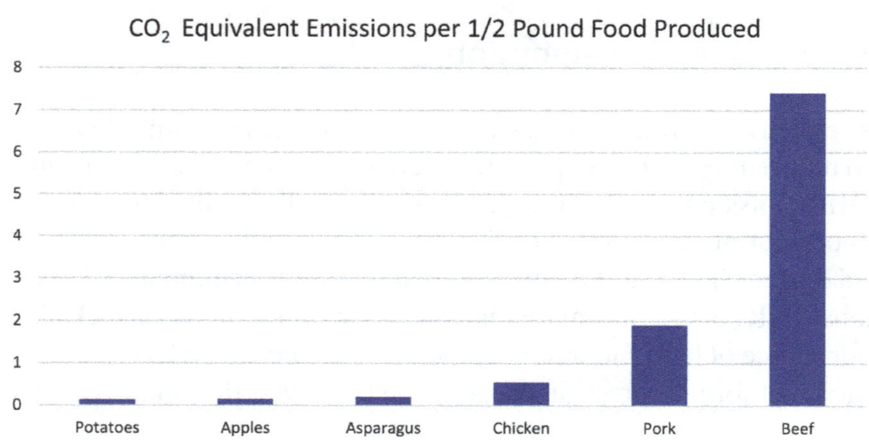

by 15 percent (Meatless Monday, 2017). Going meatless once a week helps to reduce water usage, as it takes approximately 1,850 gallons of water to produce one pound of beef, compared with about 39 gallons of water to produce a pound of vegetables (Mekonnen & Hoekstra, 2010). In addition, going meatless reduces fuel dependence, as it takes about 25 kilocalories of fossil fuel energy to produce 1 kilocalorie of meat protein, compared with 2.2 kilocalories of fossil fuel energy per 1 kilocalorie of a grain-based protein (Pimentel & Pimentel, 2003).

Adopting Sustainable Agriculture Practices

Sustainable agriculture helps to cultivate resilience to climate change (see Chapter 7). FAO lists five principles for sustainable food and agriculture:

- Improve efficiency in the use of resources;
- Take direct action to preserve, protect, and enhance natural resources;
- Agriculture that fails to protect and improve rural livelihoods, equity and social well-being is unsustainable;
- Enhance resilience of people, communities and ecosystems;
- Establish responsible and effective governance mechanisms (FAO, 2017).

The unifying concept of sustainable agricultural practices is to mimic natural ecological systems; for example, utilizing no-till farming and reducing water use with drip irrigation, rotating crops to minimize soil depletion, integrating croplands with livestock grazing, and employing ecological pest management (Natural Geographic, 2015-2017). Planting cover crops such as rye, clover or vetch after harvesting a cash crop can suppress weeds and insects, provide erosion control, and improve soil quality. Using reclaimable wastewater and other renewable resources reduces energy costs. Agroforestry techniques such as inter-planting trees with crops and growing shade-loving specialty crops help conserve soil and water, provide wildlife habitats, and increase beneficial insect populations. Sustainable agriculture protects against biodiversity loss.

Growing Power is a sustainable agriculture initiative that is the brainchild of Will Allen, who was awarded a $500,000 MacArthur Foundation "genius" grant in 2008 for his work building community food systems. Growing Power teaches farmers low-cost sustainable growing techniques such as Allen's gravity-based aquaponics system that sends water from a 10,000-gallon fish tank through a natural filtration process that adds nutrients to be used as fertilizer for salad greens, tomatoes and other crops. Growing Power collects organic waste from city businesses and converts it into nutrient-rich compost for use in raised beds that maintain their fertility for up to five years. Growing Power's Market Basket Program, a Sustainable Agriculture Research and Education (SARE) initiative, delivers up to 25 pounds of affordable produce each week to inner-city families (SARE, 2017).

REDUCING RELIANCE ON FOSSIL FUELS

Fossil fuels are nonrenewable, natural energy sources such as coal, oil, and gas, that formed in the geological past from the remains of living organisms; when fossil fuels are burned, they emit greenhouse gases that contribute to global warming. In the past two decades, the burning of fossil fuels accounted for 75 percent of human-induced GHG emissions (U.S. Department of Energy, 2017). The most consumed fossil fuel is petroleum, which can be distilled into gasoline, kerosene, naphtha, benzene, paraffin, asphalt, and other chemical regents. More than one-third of crude oil production comes from OPEC (Organization of Petroleum Exporting Countries), with six member countries in the Middle East, four in Africa, and two in South America. Two OPEC countries, Venezuela and Saudi Arabia, have the world's first and second largest reserve of petroleum; however, it is estimated that the current top producer of petroleum is Russia. The United States is the largest consumer of fossil fuels (Zhou, 2017). Reducing reliance on fossil fuels is essential in order mitigate the impact of global warming.

In the United States, 30 percent of the fossil fuel GHG comes from vehicle emissions U.S. EPA, 2017c). The use of alternative modes of transportation, engines, and fuel would greatly reduce CO_2 emissions. Alternative modes of transportation include biking, walking, and taking mass transit. Another option is telecommuting and working at home rather than driving in to an office, at least several days per week. Alternative engines for vehicles are the battery-charged motors found in hybrid and electric vehicles (EVs). EVs do not use any fossil fuel and therefore have zero emissions, do not need tune ups and oil changes, and get much better gas mileage than tradition internal combustion engine (ICE) vehicles. In 2015, Norway was far and away the world leader in EV vehicles, with a market penetration of 23.3 percent (IEA, 2016). Germany has pledged to switch over to EVs by 2030, and has begun an incentivized, evidence-based marketing campaign to promote EVs. An alternative fuel for traditional ICE vehicles is ethanol, made mostly from corn and used in place of gasoline. It began to be offered at pumps in select states in 2013. There has been some controversy about whether or not it can cause engine failure in motorcycles. Today most gasoline sold in the United States is E10, 10 percent ethanol and 90 percent gasoline, to comply with the Renewable Fuel Standards program (Urlacher, 2014).

Developing Renewable Energy Sources

Developing alternative sources of renewable energy has been facilitated by recent legislation passed in the states of California, New York, and Massachusetts, which together comprise 20 percent of the United States population. The 2016 New York Clean Energy Standard and the 2015 Senate Bill 350, signed by California Gov. Brown both set goals of 50 percent of the state's energy coming from renewable resources by 2030; the 2016 Massachusetts clean energy law sets a goal of 40 percent from renewable resources by 2030 (Ciesielski, 2017).

Wind

Both New York and Massachusetts have plans to invest heavily in wind power. There is a precedent in Europe, which supplies more than 10 gigawatts, or 10,000 megawatts, of electricity from offshore wind alone. In 2016, with a total installed capacity of 153.7 GW, European wind energy usage surpassed coal consumption and became the second largest source of power generation capacity (Wind Europe, 2017). The Long Island Power Authority (LIPA) has approved building a 90 megawatt wind farm off eastern Long Island, and 85 percent of Long Island's residents support it. This wind farm will supply electricity to about 50,000 homes. Governor Cuomo has set a goal of 2,400 megawatts of offshore wind power by 2030, enough to power more than 1 million homes (Ciesielski, 2017). Massachusetts is investing in offshore wind farms.

Solar

Solar energy is gaining great popularity worldwide for its ready availability in sunny climates, and the fact that it uses no water, which makes it attractive to arid regions that have limited freshwater supply. Most solar energy use is in the form of roof panels that are composed of smaller units of photovoltaic (PV) cells that convert sunlight into electricity. Concentrating solar thermal power (CSP) uses the solar energy to heat hot water. Both PV and CSP solar energy systems use solar collectors to harness the sun's energy.

California's renewable energy program is heavily dependent on solar energy systems. Go Solar California is a statewide program that comprises the Energy Commission's New Solar Homes Partnership and the California Public Utilities Commission's California Solar Initiative. Its statewide campaign goal is 3,000 MW of solar generating capacity with a budget of $3.35 billion (California Energy Commission, 2017). Since 2010, the world has added more solar photovoltaic (PV) capacity than in the previous 40 years (IEA, 2014). Total global capacity surpassed 150 gigawatts (GW) in early 2014. Before 2012, PV deployment was in Europe, in Germany and Italy. Since 2013, the world leaders in solar PV development has been China, followed by Japan and the United States. China is expected to continue to lead the global PV market, accounting for about 37 percent of global capacity by 2050. Manufacturing of PV systems in recent years has been concentrated in Asia, particularly in China and Chinese Taipei (IEA, 2014).

Hydropower

The most widely used water-generated power is hydroelectricity created by hydropower or the production of power through the gravitational force of falling or flowing water. Hydropower plants use the energy of falling water to generate electricity. A turbine converts the kinetic energy of falling water into mechanical energy and then a generator converts the mechanical energy into electrical energy. The Hoover dam on the Colorado River in Nevada is a famous example of a hydropower plant. In the olden days, small hydropower systems were installed in rivers or streams using a water wheel (Renewables, 2016).

In 2014, electricity production from hydroelectric sources accounted for 6.1 percent of the total electricity produced in the United States; Norway got 96 percent of its electricity from hydroelectric sources (World Bank, 2017). Paraguay and Congo got 100 percent of their electricity from hydroelectric sources in 2013; no data was available for 2014. Likewise, Nepal, Mozambique, Ethiopia, and Namibia did not have data available for 2014; however, in 2013 they got 99.7 percent, 97.7 percent, 95.6 percent, and 95.6 percent of their electricity, respectively, from hydroelectric sources. Some countries did not rely on hydroelectric sources at all, including Bahrain, Benin, Botswana, Brunei Darussalam, Denmark, Israel, Kuwait, Mongolia, Niger, Oman, Qatar, United Arab Republic, and Yemen (World Bank, 2017). It is not surprising that the Middle Eastern countries, which are largely arid, and the African countries, which are experiencing drought, would choose not to use hydropower. Unlike Norway and Sweden, which rely heavily on hydropower, Denmark is relatively flat without suitable locations for hydroelectricity plants; Denmark relies heavily on wind power.

Ocean Energy

The energy of the ocean may be harnessed from waves, tidal range (rise and fall), tidal streams, ocean (permanent) currents, temperature gradients and salinity gradients. Most ocean energy technologies are still in prototype and demonstration stages. Most of these development efforts focus on tidal stream and wave energy in open waters. Tidal barrages are being used in Korea and France to block estuaries and harness power (Renewables, 2016). Estuaries are created when a saltwater body meets a freshwater body, such as in a harbor where a river empties into the ocean. Tidal barrages change the salinity of the water; thus, some parts of the estuary become almost entirely fresh water, destroying the saltwater marsh, while other areas may become too saline to support estuarine life adapted to survival in brackish water (Clark, 2006). Tidal barrage plants do not appear to fit the criteria of sustainable development because of their adverse impacts on ecosystems. The development of renewable energy sources should be undertaken in the context of sustainable development, so as to not create further environmental degradation.

Biomass Energy

Biomass is any organic matter that can be used as an energy source. Biomass energy is renewable and has been used for about a million years, since humans figured out how to burn wood to provide heat and to cook food. Today four types of biomass are in use: wood and agricultural products, solid waste, landfill gas and biogas, and alcohol fuels (e.g., Ethanol or Biodiesel). Nearly half of the world's population uses wood-based biomass for cooking and heating. In sub-Saharan Africa, 81 percent of households rely on wood-based biomass as their primary source of energy—far more than in any other region in the world (AFREA, 2011). The biomass energy sector employs a significant workforce. However, the use of indoor cookstoves is creating serious indoor pollution problems, which are creating health problems, especially for women. Biomass burning in cookstoves emits black carbon, which is a major contributor to global warming (Cho, 2016). Sub-Saharan Africa and developing countries in Asia and Latin America emit more than three-quarters of global black carbon emissions, from the burning of wood-based biomass (wood and charcoal) (Cho, 2016). The development and distribution of cleaner, more efficient cookstoves is one strategy to reduce the harm of the biomass burning (AFREA, 2011).

In India and Pakistan, cow dung is collected and used to produce biogas, which is rich in methane and provides a cost-effective, renewable, stable source of electricity. The Kolar Biogas Project in Karnataka, India, an initiative of *myclimate* and the Indian NGO SKG Sangha, is a renewable biogas energy project that began in 2012 (*myclimate*, 2012). The household is provided with a biodigester and fuel-efficient cookstove; the biodigester is placed underground and a chute leads to it so that household members can feed it dung. The dung is composted into slurry in the biodigester; biogas builds up above the slurry and is released through the gas outlet pipe at the top when the gas burner in the household cookstove is turned on. Slurry is also sent from the biodigester through a pipe to fertilize crops (myclimate, 2012). Biofuels (e.g., ethanol, biodiesel) produced from biomass accounted for 4 percent of the world's transportation fuel in 2014. With the development of a second generation of biofuels (e.g., cellulosic-ethanol, biomass-to-liquids [BtL]-diesel and bio-synthetic gas [bio-SG]), the biofuel percentage of transportation fuel is expected to increase substantially (IEA, 2016).

Geothermal Power

Geothermal energy, heat from the earth, is clean and renewable; it comes from shallow resources such as hot water as well as deep underground resources such as molten magma. Power stations harness the geothermal energy, heat water or another fluid, and use turbines to generate electricity. Geothermal resources also are used for direct heating and cooling. Swimming pools and other public baths were the single greatest use sector, accounting for nearly 45 percent of total geothermal heat capacity in 2015. Iceland is a world leader in the use of geothermal district heating; 90 percent of its households are heated with geothermal energy (Orkustofnun, 2017). The United States also relies on geothermal energy for electricity production; it is the fourth largest source of renewable energy, after hydropower, biomass, and wind power (Renewables, 2016).

IMPROVING EFFICIENCY OF EXISTING BUILDINGS AND GOING GREEN ON NEW CONSTRUCTION

The U.S. Environmental Protection Agency recommends overhauling and retrofitting buildings' heating, ventilating and air conditioning (HVAC) systems in order to improve efficiency, assessing for compliance with ventilation and thermal comfort standards, using the most energy-efficient equipment and lighting, replacing plumbing fixtures with higher-efficiency models, installing advanced energy and water meters, retrofitting buildings and landscapes with low-impact development (LID) features, reducing irrigated landscape areas, purchasing environmentally preferable materials, using green cleaning services and integrated pest management techniques, implementing materials reduction, reuse, and recycling and composting programs, and installing renewable energy systems (U.S. EPA, 2017b). Since 2003, EPA has required all large, newly leased buildings to earn the ENERGY STAR® label within one year, or within 18 months of occupancy for new construction. For new building and construction, Leadership in Energy and Environmental Design (LEED®), a building certification program sponsored by the U.S. Green Building Council, is the benchmark for sustainable development building projects (U.S. EPA, 2017b). Projects are rated across several areas of sustainability, and then receive one of four LEED rating levels: Certified, Silver, Gold and Platinum, depending on the number of points earned. The use of green building materials and renewable energy systems is a step in the direction of the final solution to the fossil fuels problem—namely, to phase out fossil fuels and switch over to usage of alternative renewable energy sources instead.

CONCLUSION

The use of innovative technologies to facilitate sustainable development include no-till farming and drip agriculture, which conserve water and reduce carbon release. Our case examples of flood plain mapping in Tanzania and surveying earthquake damage in Ecuador demonstrate how useful technologies such as drones and GIS mapping tools are in conducting a detailed risk analysis to protect geographically vulnerable areas, and in providing a rapid, effective response after a disaster.

In sustainable development, usage of renewable energy sources will vary according to the topography and availability of resources in each region: for example, Iceland has a huge store of geothermal energy, Norway is a European leader in hydropower, Denmark uses wind power, and southern California and Florida have extensive sunshine available for solar power. Israel is a role model for sustainable water management; it reuses and reclaims most of its water and is highly efficient in water usage; its use of innovative technologies has been replicated around the world. In poor regions, such as sub-Saharan Africa, where the vast majority of the population rely on biomass for fuel and heat, new technologies focus on improving the efficiency of, and reducing the emissions from, biomass-fueled cookstoves.

On an individual level, we can all learn to reduce, reuse, and recycle. Our carbon footprints can be made smaller through the use of alternative transportation such as EVs and bicycles, by installing energy-efficient appliances, by eating less energy-intensive foods (reducing meat consumption), and by switching to renewable energy sources such as solar or geothermal energy for heating. We can buy recycled products and make sure the packaging is recyclable. We can plant pesticide-free community gardens, and give cooking classes on how to prepare meals based on unprocessed, sustainably grown foods. We can support sustainable farmers. It's the right thing to do. Your grandchildren will thank you!

REFERENCES

Africa Renewable Access Energy Program (AFREA). (2011). *Wood-based biomass energy development for sub-Saharan Africa.* Washington, D.C.: World Bank.

Al-Ansari, N., Alibraheim, N., Alsaman, M., & Knutson, S. (2014). Water supply network losses in Jordan. *Journal of Water Resource and Protection, 6*(2), 87. doi:10.4236/jwarp.2014.62013

Alvarez, L. (2014, April 14). Florida lawmakers propose a salve for ailing springs. *The New York Times.*

Bi-Moon, K. (2016, February 9). UN Secretary-General's Opening Remarks at Launch of Report for World Humanitarian Summit "One Humanity: Shared Responsibility" [As delivered]. United Nations. Retrieved from https://www.un.org/sg/en/content/sg/statement/2016-02-09/un-secretary-generals-opening-remarks-launch-report-world

California Energy Commission. (2017). California renewable energy overview and programs.

California Energy Commission. (n. d.). Renewables. Retrieved from http://www.energy.ca.gov/renewables/

Centers for Disease Control and Prevention (CDC), US Department of Health and Human Services. (2015). Planning for an emergency: strategies for identifying and engaging at-risk groups. A guidance document for Emergency Managers. Retrieved from https://www.cdc.gov/nceh/hsb/disaster/atriskguidance.pdf

Cho, R. (2016). *The Damaging Effects of Black Carbon. State of the Planet. Earth Institute*. New York: Columbia University.

Ciesielski, R. M. (2017, March 15). Another Voice: Renewable energy will benefit everyone. *Buffalo News*. Retrieved from http://buffalonews.com/2017/03/15/another-voice-renewable-energy-will-benefit-everyone/

Clark, N. (2006). Tidal Barrages and Birds. *The Ibis*, *148*, 152–157. doi:10.1111/j.1474-919X.2006.00519.x

Denchak, M. (2016). Recycling: beyond the bin. New York: National Resource Defense Council; Retrieved from https://www.nrdc.org/stories/recycling-beyond-bin

Dilley, M., Chen, R. S., Deichmann, U., Lerner-Lam, A. L., & Arnold, M. … Yetman, G. (2005). Natural disaster hotspots: A global risk analysis. The World Bank Hazard Management Unit, Washington, D.C.

Distel, O. (2015, June 27). Israel's water revolution and its overflow for the world. Times of Israel Blog. Retrieved from http://blogs.timesofisrael.com/israels-water-revolution-and-its-overflow-for-the-world/

DuPlessis, J. (2016). Case Study No. 13: Using drones to inspect post-earthquake road damage in Ecuador. FSD, Geneva, Switzerland. Retrieved from http://drones.fsd.ch/wp-content/uploads/2016/11/Drones-in-Humanitarian-Action.pdf

Englander, J. (2013/2014). *High Tide on Main Street* (2nd ed.). Boca Raton, FL: The Science Bookshelf.

Europe, W. (2017). Statistics. Brussels, Belgium: Wind Europe. Retrieved from https://windeurope.org/about-wind/statistics/

Fiala, N. (2009). The Greenhouse Hamburger. *Scientific American*, *300*(2), 72–75. doi:10.1038/scientificamerican0209-72 PMID:19186709

Food and Agriculture Organization of the United Nations (FAO). (2006). Livestock a major threat to the environment. New York: United Nations. Retrieved from http://www.fao.org/newsroom/en/news/2006/1000448/index.html

Food and Agriculture Organization of the United Nations (FAO). (2017). Sustainable food and agriculture. Retrieved from http://www.fao.org/sustainability/background/principle-5/en/

Gronewald, A. D., & Stow, C. A. (2014). Water loss from the Great Lakes. *Science*, *343*(6175), 1084–1085. doi:10.1126/science.1249978 PMID:24604186

Hazony, D. (2015, October). How Israel is Solving the Global Water Crisis. *The Tower, 31*. Retrieved from http://www.thetower.org/article/how-israel-is-solving-the-global-water-crisis/

International Energy Agency (IEA). (2014). Technology roadmap. Solar photovoltaic energy. Retrieved from http://www.iea.org/publications/freepublications/publication/TechnologyRoadmapSolarPhotovoltaicEnergy_2014edition.pdf

International Energy Agency (IEA). (2016). Global EV Outlook 2016: Beyond one million electric cars. Paris, France: OECD/IEA, Electric Vehicles Initiative. Retrieved from http://www.iea.org/publications/freepublications/publication/Global_EV_Outlook_2016.pdf

Kershner, I. (2015, June 1). Israel's water revolution. *Mosaic Magazine*. Retrieved from https://mosaicmagazine.com/picks/2015/06/israels-water-revolution/

Kharas, H., & Gertz, G. (2010). *The New Global Middle Class: A Crossover from West to East*. Washington, D.C.: Wolfensohn Center for Development at Brookings.

La Rovere, E. L., & Pereira, A. S. (2014). Brazil and climate change: a country profile. London, U.K.: SciDevNet. Retrieved from http://www.scidev.net/global/policy-brief/brazil-climate-change-a-country-profile.html

Mekonnen, M. M., & Hoekstra, A. Y. (2010). *The green, blue and grey water footprint of farm animals and animal products, Value of Water Research "Report Series No. 48"*. Delft, the Netherlands: UNESCO-IHE.

Monday, M. (2017). Why Meatless? The Monday Campaigns, Inc. Retrieved from http://www.meatlessmonday.com/contact-us/

Myclimate. (2012). From cow dung to biogas in Karnataka, India. The Climate Protection Partnership. Retrieved from https://www.myclimate.org/fileadmin/myc/klimaschutzprojekte/indien-7149/klimaschutzprojekt-indien-7149-project-story.pdf

National Capital Poison Center. (2017). Swallowed a button battery? Battery in the nose or ear? Washington, D.C.: National Capital Poison Center. Retrieved from http://www.poison.org/battery

National Geographic. (2017). Wildfires. Washington, D.C.: National Geographic Society. Retrieved from http://www.nationalgeographic.com/environment/natural-disasters/wildfires/

Nourishing the Planet. (2010). Innovation of the Week: Slow and Steady Irrigation Wins the Race. Retrieved from http://blogs.worldwatch.org/nourishingtheplanet/innovation-of-the-week-slow-and-steady-irrigation-wins-the-race-zimbabwe-africa-agriculture-farmers-water-irrigation-international-development-enterprises-ide-income-hunger-poverty-food-security-droug/

Office of the Director of National Intelligence, National Intelligence Council. (2012). *Global Trends 2025: A Transformed World.* Washington, D.C.: National Intelligence Council.

Orkustofnun. (2017). Direct use of geothermal resources. Reykjavik, Iceland: National Energy Authority of Iceland. Retrieved from http://www.nea.is/geothermal/direct-utilization/nr/91

Pimentel, D., & Pimentel, M. (2003). Sustainability of meat-based and plant-based diets and the environment. *The Journal of Clinical Nutrition*, *78*(Suppl.), 660S–663S. PMID:12936963

Renewables Ren21. (2016). Global Status Report. Retrieved from http://www.ren21.net/wp-content/uploads/2016/06/GSR_2016_Full_Report_REN21.pdf

Report, B. (1987). World Commission on Environment and Development. New York: United Nations; Available at http://www.un-documents.net/our-common-future.pdf

Scientific American. (2017). The Impacts of Relying on Desalination for Water. Retrieved from https://www.scientificamerican.com/article/the-impacts-of-relying-on-desalination/

Siegel, S. M. (2015). *Let there be water. Israel's solution for a water-starved world*. New York: St. Martin's Press.

Smart Water Networks Forum Research. (2010). *Stated NRW (Non-Revenue Water) Rates in Urban Networks*. Portsmouth, U.K.: Smart Water Networks Forum.

Smith, P., Bustamante, M., Ahammad, H., & Clark, H. Dong, H., … Tubiello, F. (2014). Agriculture, Forestry and Other Land Use (AFOLU). In O. Edenhofer, R. Pichs-Madruga, Y. Sokona et al. (Eds.), Climate Change 2014: Mitigation of Climate Change. Contribution of Working Group III to the Fifth Assessment Report of the Intergovernmental Panel on Climate Change. UK: Cambridge University Press.

Smith, P., Martino, D., Cai, Z., Gwary, D., Janzen, H., & Sirotenko, O. et al. (2007). Agriculture. In B. Metz, O.R. Davidson, P.R. Bosch et al. (Eds.), *Climate Change 2007: Mitigation. Contribution of Working Group III to the Fourth Assessment Report of the Intergovernmental Panel on Climate Change*. Cambridge, U.K. & New York, USA: Cambridge University Press.

Sustainable Agriculture Research & Education (SARE). (2017). *What is Sustainable Agriculture?* College Park, MD: SARE.

Swiss Foundation for Mine Action (FSD). (2015). *Flood Mapping for Disaster Risk Reduction: Obtaining High-Resolution Imagery to Map and Model Flood Risks in Dar es Salaam*. Geneva, Switzerland: FSD.

TeleSUR. (2016). Ecuador Earthquake Death Toll Rises to 660. Caracas, Venezuela: TeleSUR. Retrieved from http://www.telesurtv.net/english/news/Ecuador-Earthquake-Death-Toll-Rises-to-660-20160430-0025.html

The Water Project. (2017). Water conservation. Retrieved from https://thewaterproject.org/water_conservation

UNISDR. (2017). Disaster statistics. New York: United Nations. Available at: http://www.desinventar.net/index_www.html

UNISDR. (2015). Global Assessment Report on Disaster Risk Reduction. Retrieved from http://www.preventionweb.net/english/hyogo/gar/2015/en/home/GAR_2015/GAR_2015_7.html

United Nations. (2015). World population prospects. Key findings and advance tables. Retrieved from https://esa.un.org/unpd/wpp/Publications/Files/Key_Findings_WPP_2015.pdf

United Nations. (2017). Sustainable Development Goals: 17 goals to transform our world. Retrieved from http://www.un.org/sustainabledevelopment/hunger

Urlacher, R. (2014). LAB-40-The E15 Ethanol Fuel Controversy & Breakdown | Should I use E15 Fuel in my Motorcycle? (Podcast). Retrieved from https://www.lawabidingbiker.com/contact/

U.S. Department of Energy (DOE). (2017). Energy sources: fossil. Retrieved from https://energy.gov/science-innovation/energy-sources/fossil

U.S. Environmental Protection Agency (EPA). (2016a). Household Hazardous Waste. Retrieved from https://www.epa.gov/hw/household-hazardous-waste-hhw

U.S. Environmental Protection Agency (EPA). (2016b). Hazardous Waste Recycling Regulations. Retrieved from https://archive.epa.gov/epawaste/hazard/web/html/regulations.html

U.S. Environmental Protection Agency (EPA). (2017a). Electronics donation and recycling. Retrieved from https://www.epa.gov/recycle/electronics-donation-and-recycling#where

U.S. Environmental Protection Agency (EPA). (2017b). Third-party high performance building certifications at EPA. Retrieved from https://www.epa.gov/greeningepa/third-party-high-performance-building-certifications-epa

U.S. Environmental Protection Agency (EPA). (2017c). Sources of greenhouse gas emissions. Retrieved from https://www.epa.gov/ghgemissions/sources-greenhouse-gas-emissions

Wallace, S. (2017). Last of the Amazon. Washington, D.C.: National Geographic. Retrieved from http://environment.nationalgeographic.com/environment/habitats/last-of-amazon/#page=1

Wines, M. (2014, December 17). States in parched Southwest take steps to bolster Lake Mead. *The New York Times*.

World Bank. (2017). Databank: World Development Indicators. Electricity from hydroelectric sources. Retrieved from http://databank.worldbank.org/data/reports.aspx?source=2&series=EG.ELC.HYRO.ZS&country

World Health Organization. (2016). Drinking water. Fact sheet. Retrieved from http://www.who.int/mediacentre/factsheets/fs391/en/

Worldometers. (2017). Current world population. Retrieved from http://www.worldometers.info/world-population/

Zhou, P. (2017). Fossil fuels: Petroleum, coal and natural gas power much of the planet. ThoughtCo. Retrieved from https://www.thoughtco.com/where-are-fossil-fuels-found-1434558

Zwingle, E. (1993). Ogallala aquifer: Wellspring of the high plains. *National Geographic*, *184*, 80–109.

KEY TERMS AND DEFINITIONS

Chemically Reactive: Unstable and likely to explode or produce harmful fumes when exposed to other compounds.

Corrosive: In terms of waste, a substance that can corrode, or break down, metals.

Desalination: The process of removing salt from seawater to create freshwater.

Ecological Agriculture (Acroecology): An ecological approach to agriculture that views agricultural areas as ecosystems and is concerned with the ecological impact of agricultural practices.

Flammable: Likely to catch on fire.

Fossil Fuels: Nonrenewable natural fuels such as coal or gas, formed in the geological past from the remains of living organisms; when fossil fuels are burned as fuel, they emit greenhouse gases that contribute to global warming.

Gro Harlem Brundtland: A physician who was a three-time Prime Minister of Norway, Director-General of the World Health Organization, and Chair of the World Commission on Environment and Development; her 1987 report to the United Nations, known as the Brundtland Report, is a landmark in sustainable development.

Hazardous Waste: Any solid or liquid waste that is considered toxic, chemically reactive, flammable or corrosive (see toxic, chemically reactive, flammable, corrosive).

Infrastructure: The basic facilities, services, and installations needed for the functioning of a community or society, such as transportation and communications systems, water and power lines, and public institutions including schools, post offices, and prisons.

King Tides: Exceptionally high tides that occur when the sun, moon, and earth align; King Tides generally occur in coastal areas two times a year.

Recycling: The process of converting waste materials into new materials and objects.

Renewable Energy: Energy from a source that is not depleted when used, such as wind or solar power.

Simcha Blass (1897-1982): Born in Warsaw, Poland, Blass was the most influential water engineer and inventor in the Jewish settlement of Palestine; he was a major figure in the water revolution in Israel and with his son Yeshayahu invented drip irrigation, the greatest agricultural innovation of the second half of the twentieth century.

Sustainable Development: Development that meets the needs of the present without compromising the ability of future generations to meet their own needs.

Toxic: In terms of waste, it means it is harmful to human health when a person is exposed to the substance through inhalation, ingestion or touch.

Chapter 7
Cultivating Resilience

ABSTRACT

Adaptation to the challenging impacts of global warming, especially extreme weather events such as intense heat waves and hurricanes, is much more effective when the members of a community look out for each other. As we will see in the case example from the great Chicago Heat Wave of 1995, among the poor communities that were hardest hit, the neighborhoods that had social cohesion fared far better than those whose members were socially isolated. Social networking cultivates resilience, which protects vulnerable populations before, during, and after emergencies. In addition, the use of technologies is vital to address the impacts of extreme weather events; this chapter demonstrates how technologies aid in addressing heat waves and other natural hazards by providing a platform for database management, dispensing health and emergency information rapidly, and providing timely, effective medical relief.

INTRODUCTION: DEFINING SOCIAL COHESION

The term "social cohesion" has been used for decades by social scientists and sociologists. The Organisation for Economic Cooperation and Development (OECD, 2011), based in Paris, defines a cohesive society as follows:

A cohesive society works towards the well-being of all its members, fights exclusion and marginalisation, creates a sense of belonging, promotes trust, and offers its members the opportunity of upward social mobility.

DOI: 10.4018/978-1-5225-3414-3.ch007

A cohesive society works towards enhancing the well-being of all, creating a sense of belonging and promoting trust; it eschews exclusion and marginalization (OECD, 2011). Social cohesion has a synergistic effect, which makes its collective body stronger than the sum of its individual members. It provides an avenue for upward mobility. Fostering social cohesion requires that the joint stakeholders of a society—or community—work actively together to address collective action.

The OECD (2011) notes that in the decade from 2000 to 2010, there was a shift in economic growth, from West to East, and from North to South, with developing countries growing much more rapidly than in the past. Increased global mobility and migration led to increasing numbers of immigrants seeking their fortune outside their native countries. The migration problem has accelerated since 2010, due to drought, political instability, warfare, and military takeovers of governments. To facilitate the integration of immigrants into their new homes, a set of policies is needed covering employment, education, and housing, preventing discrimination and social exclusion, and facilitating entrepreneurship (OECD, 2011).

SOCIAL COHESION AS A TOOL FOR ADAPTATION TO CLIMATE CHANGE

Social cohesion is a useful tool in adaptation to climate change, as it fosters both social resilience and climate resilience. Social resilience is needed to cope with disparities in resources; when residents in a poor neighborhood work together as a collective group, in tandem with civic leaders, they develop an outreach network to inform them and help them stay safe during extreme weather events. Social resilience helps to address issues of environmental justice. Poor people usually fare the worst in natural disasters and extreme weather events, due to inferior housing, limited access to transportation, and fewer resources, including alternative places to stay. Low-income housing is particularly vulnerable in extreme weather events. Eight years after Hurricane Katrina devastated New Orleans, less than half of its public housing units had been rebuilt. The large public housing developments were torn down and redesigned as mixed-income communities; many low-income residents were pushed out of the city, as they could no longer afford the rents. Five years after Hurricane Ike destroyed 70 percent of the structures in Galveston, Texas, only 40 of the 569 public housing units lost in the storm had been rebuilt (White, 2013). Poor people often are permanently dislocated by weather disasters; their housing is not rebuilt and they have to search for affordable housing elsewhere, sometimes hundreds of miles away. The extent of the damage wrought by Hurricane Katrina was unprecedented in the United States, as it displaced New Orleans' entire population of 454,000 residents. In contrast, the previous most costly natural disaster in the United States in recent history was Hurricane Andrew in 1992; it caused the evacuation of 350,000 residents of Dade County, Florida (Sastry & Gregory, 2014). Prior to Hurricane Katrina, New Orleans had a poverty rate of 28 percent, one of the highest in the nation (U.S. Census Bureau, 2000), and a population that was 69 percent Black, with the balance White (28 percent), and Asian (2 percent). There were major disparities in return rates by race: a year after Hurricane Katrina, only 44 percent of Blacks had returned to the New Orleans metropolitan area, versus 67 percent of non-Blacks (Sastry & Gregory, 2014).

While the displacement due to Hurricane Katrina was shocking in its enormity, it is not an isolated event. As pointed out earlier, Hurricane Andrew led to massive displacement as well. In addition, over a five-year period beginning in 2010, the United States experienced extreme weather events that resulted in displacement of tens of thousands of residents in the following communities:

- In 2010, 10,000 people in Tennessee were displaced by flooding (Tennessean, 2011).
- In 2011, more than 12,000 people in Minot, North Dakota, were forced to evacuate due to flooding (Carey, 2011).
- In 2012, Superstorm Sandy displaced an estimated 10,000 people to 40,000 people in New York and New Jersey (Barron et al., 2012).
- In 2013, more than 11,000 people were evacuated due to flooding in Colorado (Rael, 2013).
- In 2014, thousands of people were forced to evacuate from wildfires in California (Whitcomb, 2014) and Washington (Atkin, 2014).

Hurricane Katrina was a textbook case of an unprepared community caught unawares. Preparation and planning are critical in adaptation to climate change. Developing a socially cohesive community prepares residents to respond more effectively when extreme weather events hit. Social cohesion creates networks—people to call in an emergency, for transportation, food, or assistance getting out of their home. Social cohesion improves a community's climate resilience, and makes evacuations safer and quicker. A chain of communication identifies each member of the community and helps them to learn and follow emergency preparedness rules. Social cohesion can greatly improve the outcome for poor neighborhoods when extreme weather events hit, as the following example demonstrates.

Case Example 1: The Chicago Heat Wave of 1995

In July, 1995, Chicago experienced an extreme heat wave that resulted in about 739 fatalities (Klinenberg, 2002). By the afternoon of the second day of the heat wave, thousands of Chicagoans had developed heat-related illnesses; paramedics could not keep up with emergency calls and hospitals were overwhelmed (Klinenberg, 2002). Twenty-three hospitals turned away ambulances, as they had no more beds. Most of these hospitals were on the South and Southwest Side of Chicago. Klinenberg (2002) reports that, in 1995, the morgue typically received about 17 bodies a day; it had a total of 222 bays. After three days of the heat wave, the morgue's capacity was exceeded by hundreds, and the county had to bring in a fleet of refrigerated trucks to store the bodies. Klinenberg (2002) called the Chicago Heat Wave of 1995 a social disaster more than a natural disaster because of the impact on low-income residents, especially isolated older adults and African Americans, who had no air conditioning and sealed themselves up in their apartments because of fear of crime:

Hundreds of Chicago residents died alone, behind locked doors and sealed windows, out of contact with friends, family, and neighbors, unassisted by public agencies or community groups. There's nothing natural about that.

The death toll was the result of distinct dangers in Chicago's social environment: an increased population of isolated seniors who live and die alone; the culture of fear that makes city dwellers reluctant to trust their neighbors or, sometimes, even leave their houses; the abandonment of neighborhoods by businesses, service providers, and most residents, leaving only the most precarious behind; and the isolation and insecurity of single room occupancy dwellings and other last-ditch low-income housing (Klinenberg, 2002 interview).

After the disastrous 1995 heat wave, the City of Chicago took steps to protect its most vulnerable residents should another extreme heat wave occur. In fact, it did, in 1999. This time, the city issued clear and emphatic warning and press releases about the impending heat wave. The city opened cooling centers and provided free bus transportation to the centers, phoned elderly residents, and sent police officers and city workers door-to-door to check up on seniors who lived alone. As a result of the aggressive response, the death rate was considerably lower than in 1995 but still catastrophic: 110 Chicagoans died in the 1999 heat wave (Klinenberg, 2002).

The 1995 Chicago heat wave exposed socioeconomic problems that contributed to the high death toll; eight of the ten neighborhoods with the highest death rates were African American with poverty and violent crime. However, three of the ten neighborhoods with the lowest death rates were also poor, violent and predominantly African American (Klinenberg, 2013). What was the key factor protecting the residents of the three neighborhoods with low death rates? *Social cohesion.* A comparison of demographically similar (low-income African American) neighborhoods showed that the ones with low death rates had more sidewalks, stores, restaurants, and community organizations that brought people together. Residents knew their neighbors; they participated in block clubs and church groups. During the heat wave, block clubs organized "wellness checks," knocking on the doors of elderly or sick neighbors to make sure that they were okay (Klinenberg, 2013). Another observation from the 1995 heat wave: women fared better than men because of their greater reliance on social networks. Social cohesion can save the lives of residents impacted by an extreme weather event. It also can lengthen life expectancy; residents in the socially cohesive low-income neighborhood had a life expectancy five years longer than those in a demographically similar adjacent neighborhood that lacked social cohesion (Klinenberg, 2013).

Social cohesion can serve as a resilience tool before, during, and after an extreme weather event (Bausson, 2015), as follows:

- Before an extreme weather event, prepare by identifying and mapping low-income, climate-vulnerable communities, and help them to develop social networks and assemble address and phone numbers of their most at-risk (i.e., elderly, sick, disabled) residents to check on during the emergency. Target the low-income, climate vulnerable communities for measures to protect them from extreme weather, such as weatherization programs, and transportation to cooling centers in heat emergencies (as was done in Chicago).
- During an extreme weather event, residents in socially cohesive communities can provide supplies, do "wellness checks," help prevent displacement, and identify local need for government officials.

- After an extreme weather event, socially cohesive communities may have a shorter duration of displacement and recover more quickly. Socially cohesive communities participating in voluntary coastal buyback programs may receive more compensation than individual residents (Bausson, 2015).

ACHIEVING SOCIAL JUSTICE AND HEALTH EQUITY

Social justice is an issue that came up after the Chicago heat wave of 1995 and that continues to surface whenever extreme weather events occur around the world. The World Health Organization (2008, p. 24) warns of the dire health consequences—including premature death—in communities with social inequity:

Social inequity manifests across various intersecting social categories such as class, education, gender, age, ethnicity, disability, and geography… People who are already disenfranchised are further disadvantaged with respect to their health – having the freedom to participate in economic, social, political, and cultural relationships has intrinsic value.

Social justice is the foundation of health equity. Life expectancy, quality of life, recovery from extreme weather events such as heat waves, floods, and storms—all are tied to access to adequate resources, including food, shelter, clean water, medicine, and health care. In 2005, the World Health Organization established the Commission on the Social Determinants of Health, a global collaborative of researchers, policymakers, and civil leaders with political, advocacy and academic experience; the goal of the Commission was to "marshal evidence on what can be done to promote health equity and to foster a global movement to achieve it" (WHO, 2008, ii). In 2008, the Commission published its findings, entitled "Closing the gap in a generation: health equity through action on the social determinants of health." The Commission noted that where people live affects their health; in 2007, for the first time, the majority of human beings worldwide were living in urban settings (WHO, 2008). Of course, there was variation from region to region. However, almost 1 billion people around the world were living in slums. Urbanization has been reshaping health problems among the urban poor, towards non-communicable diseases, accidental and violent injuries, and adverse health impacts and death from ecological disasters (WHO, 2008). The Commission's overarching recommendations are threefold:

- Improve daily living conditions–especially the well-being of girls and women and the circumstances in which their children are born, early child development and education for girls and boys, and living and working conditions.
- Tackle the inequitable distribution of power, money, and resources–such as those between men and women–by addressing the way society is organized: foster a strong public sector that is committed, capable, and adequately financed, and an accountable private sector, where society agrees on public interests and reinvests in the value of collective action.
- Measure and understand the problem and assess the impact of action–governments and international organizations, supported by WHO, should set up surveillance systems to measure and evaluate the health equity impact of policy and action. Investment in training of policy-makers and health practitioners is needed to enhance public understanding of social determinants of health, and conduct public health research and implement interventions based on social determinants of health (WHO, 2008).

The Commission stated that it believed it was possible to close the gap between the "haves" and the "have nots" in a generation by partnering with governments and civil bodies worldwide to monitor and provide assistance to geographic areas that are plagued by extreme poverty and lack of access to resources. The Commission said it is not only possible, it is imperative, to act now to take global action to achieve health equity.

It is essential that governments, civil society, WHO, and other global organizations come together in taking action to improve the lives of the world's citizens. Achieving health equity within a generation is achievable, it is the right thing to do, and now is the right time to do it. (WHO, 2008, p. ii)

The Commission concluded with an urgent moral appeal for all nations to come together in a global effort to end social injustice:

Reducing health inequities is, for the Commission on Social Determinants of Health, an ethical imperative. Social injustice is killing people on a grand scale. (WHO, 2008, p. 36)

The Social Determinants of Health

The social determinants of health are a leading health indicator in the United States' national blueprint for health, *Healthy People 2020,* which provides evidence-based 10-year objectives aimed at improving the health of all Americans (*Healthy People 2020*, 2017). *Healthy People 2020's* objectives were developed by the U.S. Department of Health and Human Services Secretary's Advisory Committee, a 12-member body comprising nationally recognized experts in public health. *Healthy People 2020* draws on the four previous *Healthy People* initiatives:

- **1979 Surgeon General's Report:** *Healthy People: The Surgeon General's Report on Health Promotion and Disease Prevention;*
- ***Healthy People 1990:*** *Promoting Health/Preventing Disease: Objectives for the Nation;*
- ***Healthy People 2000:*** *National Health Promotion and Disease Prevention Objectives;*
- ***Healthy People 2010:*** *Objectives for Improving Health.*

Healthy People 2020 has four overarching goals:

- Attain high-quality, longer lives free of preventable disease, disability, injury, and premature death;
- Achieve health equity, eliminate disparities, and improve the health of all groups;
- Create social and physical environments that promote good health for all;
- Promote quality of life, healthy development, and healthy behaviors across all life stages.

Healthy People 2020 drew on the World Health Organization's (2008) publication on the importance of the social determinants of health to longevity, quality of life, and social justice. Both WHO and *Healthy People 2020* advocate using the social determinants of health as a framework for addressing and eliminating health disparities, creating a socially cohesive, resilient society in which community and the neighborhood are key supportive elements. *Healthy People 2020* lists five key components of the social determinants of health (SDOH), as shown in Figure 1.

The five components are depicted as an interconnected ring because they are mutually interdependent on each other. For example, inadequate healthcare such as management of childhood asthma can lead to excessive school absenteeism and impede a child's educational progress. Likewise, economic instability may cause a family to become homeless, moving from relative to relative to shelter, jeopardizing the children's education and undermining the development of protective social networks. The great 19th century school reformer Horace Mann (1843) summed up the importance of health to education: "On the broad and firm foundation of health alone can the loftiest and most enduring structures of the intellect be reared" (p. 532).

In short, Mann (1843) warned the public that the entire structure of the American educational system rested on a foundation of good health. He knew very well what it was like to suffer from poverty and poor health, as he had developed health problems from going to work as a manual laborer at a young age. In Mann's day, the American life expectancy was less than 50

Figure 1. Five key components of the social determinants of health (SDOH)
Source: Healthy People 2020 (2017).

years, and penicillin had not yet been discovered. The leading causes of death were infectious diseases such as pneumonia and influenza. Today, the life expectancy in the United States is close to 80 years, and the leading causes of death are chronic diseases such as heart disease, cancer, and chronic lower respiratory diseases (Arias, 2014; Miniño et al., 2011).

Critical Shortages of Doctors, Nurses, and Midwives

Today several African countries are still suffering from such dire poverty and poor health that life expectancy is around 50 years of age, where it was in the United States in the 19th century (WHO, 2008). Contributing to the problem is a critical shortage of trained healthcare workers in sub-Saharan Africa, which had only 4 percent of health workers but 25 percent of the global burden of disease. For the same period, the Americas had 37 percent of health workers but only 10 percent of the global burden of disease. The World Health Organization (2006) estimated that there was a worldwide shortage of almost 4.3 million doctors, midwives, nurses, and support workers, with critical shortages in Africa, the Eastern Mediterranean, and Southeast Asia (WHO, 2006). There is a direct correlation between human resources for health (HRH) density (per 10,000 population) and infant, child, and maternal survival. An HRH density of 22.8 per 10,000 was estimated as the minimum threshold. In 2010, the World Health Organization surveyed the 49 countries with the most critical shortages of doctors, nurses, and midwives. Only five of the 49 were above the minimum threshold of 22.8 per 10,000 population. Shockingly, the bottom five countries (10.2 percent of the 49 countries) had a density of only 1 or 2 doctors, nurses and midwives per 10,000 population. Another 26.5 percent had a density of only 3-5 doctors, nurses and midwives per 10,000 population. Figure 2 shows the density of doctors, nurses, and midwives going clockwise from a healthy green (5 countries, or 10.2 percent of the 49 countries above the threshold of 23) to virtually no clinicians (5 countries, or 10.2 percent).

Millennium Development Goals

In 2000, all 189 member states of the United Nations signed the Millennium Declaration, which was committed to achieving the following eight Millennium Development Goals (MDGs) by 2015:

Figure 2. Density of doctors, nurses and midwives in 49 priority countries
Source: WHO Global Atlas of the Health Workforce (2010)

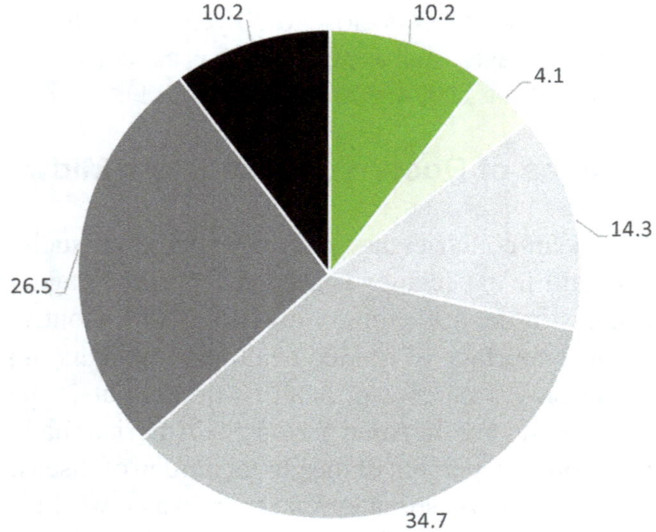

Goal 1: Eradicate extreme poverty and hunger.
Goal 2: Achieve universal primary education (for boys and girls).
Goal 3: Promote gender equality and empower women.
Goal 4: Reduce infant/child mortality.
Goal 5: Improve maternal health.
Goal 6: Combat HIVAIDS, malaria, and other diseases.
Goal 7: Ensure environmental sustainability.
Goal 8: Develop a global partnership for development (United Nations, 2015a).

The Millennium Declaration (United Nations, 2015a) and its eight MDGs laid out a course of action and funded a number of international initiatives to address specific inequities. A major initiative for maternal and child health was the *Muskoka Initiative,* launched in Canada in 2010; its aim was to significantly reduce infant and child mortality under age five as well as maternal death in pregnancy and childbirth. Its efforts were focused on the 10 countries where infant/child and maternal mortality rates were highest—Afghanistan, Bangladesh, Ethiopia, Haiti, Malawi, Mali, Mozambique, Nigeria, South Sudan, and Tanzania (Muskoka Initiative, 2014). The Muskoka Initiative ended in 2015, but Canada committed another 3.5 billion for the next five years to improve infant, child, and maternal health. Reduction in maternal and child mortality is tied to access to resources and technology, specifically: 1) training and deployment of an adequate healthcare workforce in nations with severe shortages of such workers, 2) better registration, electronic record keeping, and data collection so that statistics on mortality, disease, and other health items are accurate and complete, 3) provision of prenatal vitamins to pregnant women through the Micronutrient Initiative, 4) provision of vaccinations through the Vaccine Alliance, and Tetanus shots through the Eliminate Maternal and Neonatal Tetanus Initiative (Muskoka Initiative, 2014).

Every Woman, Every Child is an initiative that addresses the gender inequity faced by women and girls in many countries around the world, including equal access to education. Launched by the United Nations in 2010, *Every Woman, Every Child* mobilizes and intensifies action by governments, multilaterals, the private sector and civil society to address major health challenges facing women, children and adolescents around the world.

Tarja Halonen, President of Finland from 2000 to 2012, and Michelle Bachelet Jeria, President of Chile from 2006-2010 and 2014-2018, are members of the High Level Advisory Committee of *Every Woman Every Child.* Both are role models and tireless advocates for gender equity. Both women know a thing or two about breaking glass ceilings, as they both were the first female Presidents of their countries. In their article, "Breaking the ceiling for a better future," Bachelet Jeria and Halonen (2016) note that, compared to men and boys, women and girls often experience reduced access to education, information, technology, and health care services. They emphasize the importance of education for girls; worldwide, 62 million girls between the ages of six and 15 are not in school, and girls continue to lag substantially behind boys in secondary school completion rates. Bachelet Jeria and Halonen (2016) state that gender equity is both a goal and part of the solution:

Think of what we can collectively achieve if women and girls are empowered, included, listened to. We are both examples of what women can accomplish when they have a healthy upbringing, quality education and are able to engage in civic participation. We are now trying to make a difference in the world, so that other women and girls can overcome barriers and beat the odds to reach their full potential. Human rights, including gender equity, are at the basis of development and precursors for health and progress. Empowered women can fight for their rights and make better, more informed decisions, especially those that affect their health and wellbeing, which in turn have an impact in their families and communities. This has the potential to create a virtuous cycle of prosperity. (Bachelet Jeria & Halonen, 2016)

Freerice is an Internet-based initiative of the United Nations World Food Programme that has two goals:

- Provide education to everyone for free.
- Help end world hunger by providing rice to hungry people for free.

John Breen founded *Freerice* in 2007 and donated it to the World Food Programme (WFP) in March 2009. Its non-profit website, which is available in English, French, Italian, Spanish and Korean, offers multiple interactive games with different subjects and levels; each time a player gets a correct answer, 10 grains of rice are donated through the WFP to end hunger. The website is educational and includes a section on how to join in the worldwide effort to end hunger. The rice is distributed to countries such as Cambodia, Myanmar, Bangladesh, Uganda, Nepal, Bhutan, and Haiti (*Freerice*, 2007-2017).

Another WFP initiative, *Food Assistance for Assets (FFA)*, has the following aims:

- To improve access to food for the most vulnerable and food-insecure people;
- To boost access to livelihood assets that ensure early recovery, reduce disaster risks and build long-term resilience to shocks;
- To promote gender equality (FFA, 2017).

FFA provides a sustainable food program; in exchange for food vouchers, communities participate in work that increases their food security and resilience in the face of natural disasters. Projects include repairing irrigation systems, building bridges, implementing soil conservation, and setting up community granaries. *FFA* addresses the underlying issues of food insecurity to ensure adequate food supply year-round. It works on natural resource development and management (soil and water conservation, flood control, water harvesting) and supports restoration of agricultural, pastoral and fisheries potential (*FFA*, 2017). *FFA* also repairs and constructs public projects such as roads and bridges, and community infrastructures such as schools, marketplaces, and latrines. *FFA* provides skills training to manage assets and generate income, and helps communities to access insurance. In 2014, 12.7 million people in

52 countries benefited from *FFA*; they rehabilitated 254,000 hectares of land and planted 14,000 hectares of forest and built 3,410 water ponds, shallow wells, and fish ponds. More than a quarter million people received trainings on managing natural resources and other assets, and on Income Generating Activities (IGAs) (*FFA*, 2017).

Vector-borne diseases account for about 17 percent of all infectious diseases worldwide, resulting in more than 1 million deaths annually. Malaria causes more than 400,000 deaths every year, with most of the fatalities in children under age five (WHO, 2016a). World Health Organization's *Global Malaria Programme* (2016b) has distributed insecticide treated nets (ITNs) free to countries where malaria case rates have been high. Vector control is the primary way to prevent malaria transmission. Two forms of vector control are in wide use: ITNs and indoor residual spraying (IRS). The World Health Organization calls ITNs "the cornerstone of malaria prevention efforts, particularly in sub-Saharan Africa." More than 900 million ITNs were distributed in sub-Saharan Africa between 2004 and 2014. Mosquito nets have been found to be very effective in combatting malaria, as they protect sleeping people from the bite of the *Anepholes* mosquito, which is active primarily during the night. The global incidence rate of malaria fell as estimated 37 percent, and the mortality rate fell by 58 percent, during the period 2000-2015 (United Nations, 2015a).

In 2015, the United Nations issued a Millennium Development Goals Report (United Nations, 2015a) assessing how much progress had been made toward achieving the eight MDGs. The report found that, overall, the 15-year effort had been successful in reducing poverty, improving educational enrollment across the globe, and empowering women and girls. Globally, the number of people living in extreme poverty decreased substantially, from 1.9 billion in 1990 to 836 million in 2015, with most of the progress occurring after 2000. In the developing regions, the proportion of undernourished people decreased from 23.3 percent in 1992-93 to 12.9 percent in 2014-16. In 2015, more girls were enrolled in school, and women had gained ground in parliamentary representation in nearly 90 percent of the 174 countries with data. The rate of children dying before their fifth birthday decreased from 90 to 43 deaths per 1,000 live births since 1990. The maternal mortality ratio showed a decline of 45 percent worldwide, with most of the reduction occurring since 2000. From 2000 to 2013, new HIV infections fell by 40 percent, from an estimated 3.5 million cases to 2.1 million cases; in 2014, 13.6 million people infected with HIV were receiving antiretroviral therapy (ART), compared to only

800,000 in 2003. The global malarial incidence rate fell 37 percent and the mortality rate by 58 percent (United Nations, 2015a).

Technology is key to achieving the world's development goals. Mobile-cellular towers have spread rapidly; by 2015, 95 percent of the world's population was covered by a mobile-cellular signal. The number of cell phone subscriptions grew almost tenfold, from 738 million in 2000 to 7 billion in 2015. Internet penetration has increased from 6 percent of the world's population in 2000 to 43 percent in 2015 (United Nations, 2015a). Internet penetration is a boon on several fronts, as it provides: 1) a platform and tools to measure, collect, and evaluate health statistics (it is difficult to measure maternal and infant health in regions that do not have Internet access or even an adequate system for registering all births and deaths), 2) faster, more effective medical response, especially in rural areas, 3) the means to dispense health and emergency information rapidly among a population, which can be critical in the case of an epidemic or impending natural disaster, 4) an inexpensive and efficient means of training large numbers of health care workers to address critical shortages around the world, 4) educational opportunities for many who cannot afford to travel great distances to universities, and 5) a great tool for public health campaigns. Prevention is more effective and much less costly than treatment; used in connection with other media, the Internet is an ideal venue for public health campaigns:

Public health campaigns are used for promoting good hygiene and healthy choices in diet and exercise, preventing infectious diseases from spreading, managing chronic diseases such as diabetes and asthma, emergency preparedness, and getting individuals to see a healthcare provider for checkups and screenings, among other things. These campaigns are initiated using social media, television, billboards, and print media such as posters, flyers, and magazine ads. (Weiss, 2016)

Well-designed public health media campaigns use Internet (social media), TV, and radio; they are multilevel and multifaceted. The United States has used such campaigns to modify unhealthy behaviors and to promote healthy behaviors. A good example of a successful, well-designed public health campaign was the *Tips from Former Smokers*, a national government-funded anti-smoking campaign launched by the Centers for Disease Control and Promotion in 2012 (McAfee et al., 2013). The campaign ran on TV stations, aired on the radio, and was advertised in print, on billboards, and on the Internet. A survey was sent to adult smokers and non-smokers before the *Tips* campaign began and again after 12 weeks of exposure to the *Tips* campaign. Seventy percent of invited smokers, or 4,108 adults, and 58 percent of invited non-smokers, or 3,000 adults, participated in the baseline survey. Participation in the post-*Tips* survey was 74 percent for both groups, totalling 3,051 smokers and 2,220 non-smokers. *Tips* reached nearly 80 percent of U.S. smokers and had a significant positive effect: the campaign was associated with a 12 percent relative increase in quit attempts and an absolute increase in quit attempts of 3.7 percent. The government invested in the *Tips* campaign in the hopes of having a significant impact on tobacco use, which is the leading preventable cause of death in the United States (McAfee et al., 2013).

A systematic review (Schiavo et al., 2014) of public health campaigns in middle- and low-income countries (LMIC) looked at evidence on interventions that communicated risk and promoted disease mitigation measures in geographic areas where there was an epidemic or emerging disease outbreak. The reviewers found a scarcity of well-designed interventions; most were one-pronged and one-level rather than multifaceted and multilevel. They concluded that the delivery of the public health campaigns was hampered by a lack of technology and sparse resources. The consequences of low-tech, resource-poor, underfunded public health campaigns can be devastating:

The 2002-2003 SARS epidemic from China, the 2009 H1N1 pandemic, and the 2014-15 Ebola outbreak in Africa all point to the need for improving health risk communication... to help prevent or control epidemics and disease outbreaks. (Weiss, 2016)

Post-2015 Agenda: Sustainable Development Goals

In August 2015, the United Nations convened to sign an updated version of the Millennium Declaration, the Post-2015 Development Agenda (United Nations, 2015b). It changed the name of the goals from MDGs to Sustainable Development Goals (SDGs), and expanded their number to 17. Secretary General of the United Nations Ban Ki-Moon (Ki-Moon, 2015) noted that, although progress had been made toward achieving the MDGs, there was still a great deal to do:

Yet for all the remarkable gains, I am keenly aware that inequalities persist and that progress has been uneven. The world's poor remain overwhelmingly concentrated in some parts of the world. In 2011, nearly 60 percent of the world's one billion extremely poor people lived in just five countries. Too many women continue to die during pregnancy or from childbirth-related complications. Progress tends to bypass women and those who are lowest on the economic ladder or are disadvantaged because of their age, disability, or ethnicity. Disparities between rural and urban areas remain pronounced.

The 17 SDGs of the Post-2015 Development Agenda, to be achieved by 2030, are as follows:

Goal 1: End poverty in all its forms everywhere.
Goal 2: End hunger, achieve food security and improved nutrition and promote sustainable agriculture.
Goal 3: Ensure healthy lives and promote well-being for all at all ages.
Goal 4: Ensure inclusive and equitable quality education and promote lifelong learning opportunities for all.
Goal 5: Achieve gender equality and empower all women and girls.
Goal 6: Ensure availability and sustainable management of water and sanitation for all.
Goal 7: Ensure access to affordable, reliable, sustainable and modern energy for all.
Goal 8: Promote sustained, inclusive and sustainable economic growth, full and productive employment and decent work for all.
Goal 9: Build resilient infrastructure, promote inclusive and sustainable industrialization and foster innovation.
Goal 10: Reduce inequality within and among countries.
Goal 11: Make cities and human settlements inclusive, safe, resilient and sustainable.
Goal 12: Ensure sustainable consumption and production patterns.
Goal 13: Take urgent action to combat climate change and its impacts.*
Goal 14: Conserve and sustainably use the oceans, seas and marine resources for sustainable development.
Goal 15: Protect, restore and promote sustainable use of terrestrial ecosystems, sustainably manage forests, combat desertification, and halt and reverse land degradation and halt biodiversity loss.
Goal 16: Promote peaceful and inclusive societies for sustainable development, provide access to justice for all and build effective, accountable and inclusive institutions at all levels.
Goal 17: Strengthen the means of implementation and revitalize the Global Partnership for Sustainable Development.

* Acknowledging that the United Nations Framework Convention on Climate Change is the primary international, intergovernmental forum for negotiating the global response to climate change (United Nations, 2015b).

The SDGs expanded on the MDGs in several ways: first, poverty and hunger were separated into two goals, with the achievement of food security, improved nutrition, and improved sustainable agriculture added to the "end hunger" goal. This acknowledges the importance not only of consuming enough calories, but of consuming a nutritious diet on a regular basis without fear of starvation. When a population lives in fear of starvation, they are in survival mode, without the means to rise above bare subsistence, and their food insecurity causes a host of unhealthy activities, including eating unhealthy or spoiled food as well as non-food items to fill the stomach, resulting in malnourishment, sickness, and premature death. Food insecurity is often associated with the poorest nations in the world, but in the United States one in five families with children suffers from food insecurity. A recent study of food insecurity in the United States (Denney et al., 2017) investigated whether social cohesion influences food insecurity and whether it moderates the relationship between race/ethnicity and food insecurity. The researchers found that lower levels of perceived neighborhood social cohesion were associated with higher levels of food insecurity even after adjusting for household socioeconomic factors; for example, for immigrant Latina mothers the probability of food insecurity was nearly .40 percent in neighborhoods where they perceived there to be little to no social cohesion, but only .10 percent in neighborhoods where they perceived there to be high social cohesion (Denney et al., 2017). The researchers concluded that perceived social cohesion is protective against food insecurity in households with children, especially for racial and ethnic minority households at risk for food insecurity. They recommended developing programs that promote social cohesion by building more closely knit communities as a key strategy for reducing food insecurity in general, and disparities in food insecurity by race and ethnicity (Denney et al., 2017).

The most notable change from the MDGs to the SDGs of the Post-2015 Development Agenda (United Nations, 2015b) is the sustainability theme, which is found in 11 of the 17 goals. The Annex title--*Transforming our World: the 2030 Agenda for Sustainable Development*—sums it up. The 2030 Agenda is transformational – the scale and ambition are greater than in the 2015 Agenda. The 2030 Agenda seeks to build on the MDGs

...to realize the human rights of all and to achieve gender equality and the empowerment of all women and girls....and balance the three dimensions of sustainable development: the economic, social and environmental (United Nations, 2015b).

In the 2030 Agenda, sustainable agriculture was added to the SDG goal to end hunger, because sustainable agriculture is necessary to provide food security. The principle behind sustainable agriculture is that, in meeting the needs of the present, we must not compromise our ability to meet the needs of future generations (Brodt et al., 2011). After World War II, modern farming incorporated new chemical pesticides and fertilizer, mechanization, monoculture, and biotechnology to kill pests and increase food production. With government subsidies to boot, farmers were able to produce large quantities of food more rapidly (National Geographic, 2015-2017). However, there have been steep ecological and social costs:

Prominent among these are topsoil depletion, groundwater contamination, air pollution, greenhouse gas emissions, the decline of family farms, neglect of the living and working conditions of farm laborers, new threats to human health and safety due to the spread of new pathogens, economic concentration in food and agricultural industries, and disintegration of rural communities. (Brodt et al., 2011, p. 1)

Sustainable agriculture uses a variety of techniques; the common thread among them is that the farming mimics natural ecological practices; for example, minimizing tilling and water use, rotating crops to minimize soil depletion, integrating croplands with livestock grazing, and using ecological pest management (Natural Geographic, 2015-2017). Sustainable agriculture encompasses three broad goals, also known as the 3 Pillars of Sustainability (Sustainable Agriculture Research & Education, 2017):

- Profit (over the long term);
- Stewardship of our environment and natural resources;
- Quality of Life (Sustainable Agriculture Research & Education, 2017).

The 3 Pillars of Sustainability are shown in Figure 3.

Quality of life includes working and living conditions of laborers and livestock, the needs of rural communities, and consumer health and safety both in the present and the future. Stewardship of our environment and natural

Figure 3. Three pillars of sustainability
Source: Adapted from Brodt et al., 2011.

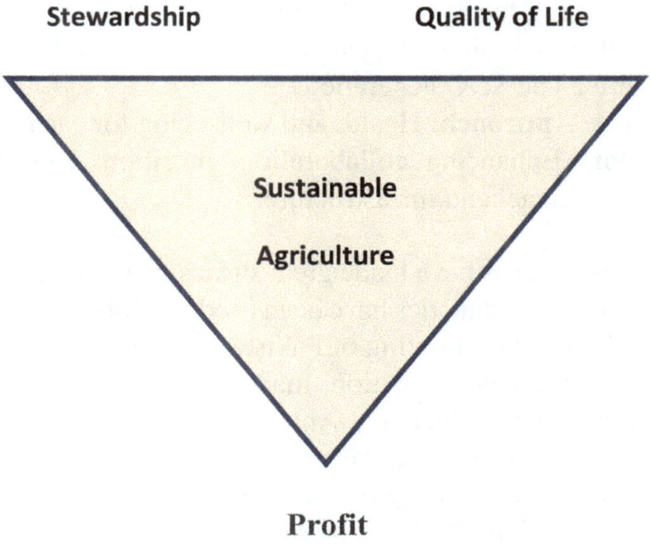

resources entails using these resources without damaging their quality and in ways that allow them to be renewed for future generations. A systems perspective is needed to understand the concept of sustainability. Food systems include food production, distribution, and consumption systems. Using a systems perspective gives us a comprehensive view of food production and distribution, as well as agriculture's impact on society and the environment. A system's survival is a function of its resilience, adaptivity, and diversity (Brodt et al., 2011).

Based on the United Nations' Post-2015 Development Agenda, the *Global Strategy for Women's, Children's and Adolescents' Health* (WHO, 2017) was updated. The *Global Strategy* provides a roadmap for ensuring the well-being of women and girls and for ending all preventable deaths of women, children and adolescents by 2030. The *Global Strategy* is fully aligned with the Sustainable Development Goals (SDGs). It (WHO, 2017) sets out three objectives to be achieved by 2030:

- **Survive:** *End preventable deaths.*
- **Thrive:** *Ensure health and well-being.*
- **Transform:** *Expand enabling environments.*

The updated *Global Strategy* (WHO, 2017) includes five important foci:

- **Equity:** A strong focus on reaching the most vulnerable.
- **Universality:** Including a focus on humanitarian and fragile settings.
- **Adolescents:** The SDG generation.
- **Life-Course Approach:** Health and well-being for each stage of life.
- **Multisector:** Enhancing collaboration: nutrition, education, water, sanitation, hygiene, and infrastructure.

In recent years, as girls have made great strides in attaining an education, highly publicized terrorist tactics have been used to frighten them away from schools; the 2012 Taliban shooting of Pakistani teenager Malala Yousafzai, who advocated for female education, made her an even more passionate spokeswoman for female education. Another example is the kidnapping of 276 female students from the Government Secondary School in Chibok, Nigeria in 2014; some of the girls escaped but others are still missing. *Every Woman, Every Child* will not be stopped by such scare tactics; it will continue to advocate for all girls' rights to an education. These terrorist acts are vicious attempts to deter girls and women from continuing on their path to gender equality and health equity. Social cohesion among women and girls will help them to continue on their path toward equal access to education, work, and healthcare.

CULTIVATING RESILIENCE

In *Resilience Thinking*, Walker and Salt (2006, p.xiii) define resilience as "the capacity of a system to absorb disturbance and still retain its basic function and structure." Resilience thinking challenges our old-fashioned notion of business as usual, which was product-oriented: create a cheaper product more efficiently and you will make a bundle! However, our world has changed; natural resources are shrinking and becoming degraded while our global population is increasing. Climate change is accelerating at a pace that will require massive relocation in the not-too-distant future. In such a world, we have a "new normal." To adapt successfully to this new normal, we need a new mindset. A *resilient mindset* challenges us to think in terms of systems, not products. The challenge is to manage our current systems with fewer resources while sustaining enough for our future needs.

Hine and his colleagues (2016, p. 142) eloquently describe how resilience takes the risk-reduction strategies of adaptation one step further.

It focuses on creative adaptation that builds a community's underlying adaptive capacity while being proactive about future hazards. Resilience addresses the ability of the social, environmental, and economic components of a community to absorb change...and still retain their basic structure and ability to function, while at the same time enhancing their capacity to rebound from unforeseen or extreme disturbances.

Case Example 2: Flooding in Boulder in 1976

The City of Boulder, Colorado's response to flooding is a good example of community resilience. Boulder is situated at the mouth of the Big Thompson Canyon, which has many streams in it that can flood due to melting snow and heavy rain. In 1976, a massive flood killed 144 people and caused extensive property damage. The following steps were taken to prevent such a large-scale disaster from happening in the future:

- The city bought and removed many buildings from flood-prone areas.
- Automatic floodgates were installed to protect buildings that could not be relocated.
- Culverts (open drains under roadways) and underpasses were constructed to divert water.
- Stream banks were restored and stabilized.
- The city elevated a major highway and improved its storm sewers.

In 2013 there were torrential rains in Boulder and the surrounding region, resulting in massive flooding. The city was able to maintain critical services and property damage was less extensive than in 1976. There were seven deaths, compared to 144 in the flood of 1976. The city recovered more quickly than it had in the past; its resilience planning had enhanced its capacity to rebound from an extreme weather event. After the 2013 flood, the city took additional measures to flood-proof crucial facilities. A community's resilience is an ongoing process of creative adaptation to the demands of the environment, including extreme weather events (Hines et al., 2014).

Empowering Communities

Every city should have a comprehensive plan that includes emergency preparedness, recovery, and mitigation. Coastal cities and others vulnerable to flooding need to approach their planning with the understanding that flooding will be increasingly frequent and severe due to climate change. Buildings in coastal areas should have setbacks. Policies should be implemented to discourage development of new infrastructures in vulnerable coastal areas. That is no easy task, as beachfront property has great appeal to buyers, who pay a hefty price for the ocean view and beach access.

Case Example 3: Severe Storm Damage in New York in 2012

After a coastal area has been impacted by a storm or flood, the state may need step in to help relocate people. An example is the NY Rising Community Reconstruction Program (NYRCRP, 2016), which was created after three severe storms (Hurricane Irene, Tropical Storm Lee, Superstorm Sandy) battered New York in one year. Superstorm Sandy left hundreds of thousands of households without power, caused sewage to contaminate waterways, washed away protective coastal barriers, and closed businesses for weeks. Recognizing the need for a statewide agency to deal with the disaster's aftermath, in 2013 Governor Cuomo established the Governor's Office of Storm Recovery (GOSR), under the umbrella of NY Rising, with a $700 million grant; by late 2016, a total of $17 billion had been committed to repairing homes and businesses, restoring community services, and mitigating future storm risks across New York State (NYRCRP, 2016). Funding sources include HUD's Community Development Block Grant—Disaster Relief Program (CDBG-DR), FEMA's Hazard Mitigation Grant Program (HMGP), and the U.S. Environmental Protection Agency (EPA). Some of the funds were used to create the NY Rising Acquisition Program, which offered buyouts and resettlement incentives to those in the most impacted areas.

NYRCRP (2016) is a participatory recovery and resilience initiative. Unlike traditional top-down approaches to disaster recovery, NYRCRP engages residents, business owners, and civic leaders in a democratic, bottom-up approach. More than 650 New Yorkers have represented their communities on NYRCRP planning committees, working with the state's team to rebuild their communities and make them more resilient. The NYRCRP's participants have included traditionally underrepresented communities such as immigrants and students (NYRCRP, 2016). Each of the 66 Planning Areas (serving a total of 124 communities) had to develop a NY Rising Community Reconstruction Plan. Critical issues were identified; for example, in Greater Bay Shore, Long Island, the Planning Committee identified the following key issues (NYRCRP, 2016):

- Repetitive and severe flooding due to tidal surges, heavy rains, and extreme high tides;
- Extended recovery time for basic services;
- Resiliency of ferry terminals/marina;
- Presence of vulnerable populations;
- Communication and educational outreach before, during, and after the storm;
- Maintaining critical services and access at local hospital;
- Water quality;
- Continued recovery of the local economy including downtown area.

The Planning Committee held public meetings and solicited input from the public, including civic, medical, and religious organizations. The end result was a blueprint for recovery (NYRCRP, 2016) that offered eight strategies to address the eight critical issues that had been identified:

- Ensure adequate resources to enhance the quality of Community-Based Organizations (CBOs) to prepare for, and respond to, emergencies; add seven generators to crucial facilities.
- Ensure public safety and the ability of first responders to react promptly and effectively to severe storm events and other emergencies by upgrading communications and GIS systems and purchasing search and rescue equipment.
- Enhance the economic resilience of the community by implementing "Complete Street" improvements to parallel access and evacuation along the waterfront.
- Provide for the unique needs of vulnerable populations (e.g., the elderly, people with disabilities, low-income residents) by creating an Emergency Action and Education Plan and installing "green" drainage improvements on multi-family residential housing unit.

- Encourage safe and resilient housing for all residents by implementing housing resiliency enhancements.
- Ensure continuity of services and access to hospital and public safety services.
- Integrate "Green" and "Gray" infrastructure (natural and engineered stormwater management systems) to holistically manage stormwater and reduce flooding.
- Mitigate flooding from the bay by replacing damaged bulkheading.

NYRCRP is a huge and complex initiative spread over a large geographic area. It has been implemented in Round I (2013-2014) and Round II (2015-2016) and was still active in late 2016. The three storms—Hurricane Irene, Tropical Storm Lee, and Superstorm Sandy— caused enormous damage to a wide area in upstate NY, New York City, and Long Island. Superstorm Sandy caused $50 billion in damages; 157,165 housing units were damaged and 300,000 businesses were impacted (NYRCRP, 2016). An estimated 43,499 public infrastructure repairs and mitigation projects are needed as a result of the three severe storms. The estimated cost to repair damages is $15,045,010,000, with an additional $2,562,010,000 in mitigation costs to protect vulnerable areas from future storms and ensure that essential services such as power, water, roads and transportation recover more rapidly than in 2012 (NYRCRP, 2016).

Given the scope and severity of the problems facing residents and businesses impacted by Superstorm Sandy, as well as Hurricane Irene and Tropical Storm Lee, a new approach to recovery was needed. As one of the Planning Committees put it, creating infrastructure projects was only part of the solution; the city also needed to create a resilient mindset (NYRCRP, 2016)! The Planning Committee was right--we are living in a world whose climate is changing rapidly, with rising sea level, increasing temperatures, and increasing severity and frequency of superstorms and other extreme weather events. After New York was devastated by three such events in a year, residents of the state realized that we are no longer talking about a rare "100-year storm." The "100-year storm" has become a regular occurrence.

Individuals are encouraged to join local groups and promote community resilience through climate advocacy. Grassroots initiatives have a ripple effect, gradually pushing communities, cities, states--and eventually, the nation--on a path toward greater climate resilience. Numbers do make a difference, as we saw on April 29, 2017, when hundreds of thousands of people marched in the People's Climate March in Washington, D.C. and in sister marches all around the country. To learn more about how to become a climate advocate, contact Citizens Climate Lobby, a national nonprofit organization that has local branches around the country (https://citizensclimatelobby.org/) or call 619-437-7142.

Local and Regional Policymaking

Portnoy and Berry (2014) looked at the role nonprofit environmental groups play in local policy making. Increasingly, cities are accommodating the growing environmentalism by adopting Climate Action Plans and hiring Chief Resilience Officers and Directors of Sustainability, positions that did not exist a few years ago. In the past decade, many cities have signed initiatives aimed toward reducing greenhouse gas (GHG) emissions; prime examples are the Climate Protection Programme of ICLEI—Local Governments for Sustainability (an outgrowth of the Agenda 21 process), the Climate Protection Agreement of the U.S. Conference of Mayors, and the Climate Initiative of the Clinton Foundation. Environmentalism can be an inherent part of a city's appeal; green spaces, bike paths, and accessible, affordable mass transit enhance quality of life. Environmental advocacy seems to have come of age and is playing a key role in city policymaking. Portnoy and Berry (2014, p. 395) concluded that, based on the evidence, "inclusion of environmental groups in city policymaking is strongly linked to administrators' perceptions of city commitment to environmental protection."

Sea level rise is a pressing issue for coastal communities, and it is vital that they join forces to create adaptation plans and climate resilience. The Southeast Florida Regional Climate Change Compact (The Compact) is such a partnership; in 2009, four southeastern coastal Florida counties—Broward, Miami-Dade, Monroe, and Palm Beach—joined together to work collaboratively on climate change mitigation and adaptation strategies. The Compact has been working on changing Florida planning legislation to incorporate and fund efforts to address climate change and sea level rise (South Florida Regional Planning Council, 2013) This is no small task in a state where the governor does not acknowledge that climate change is a problem.

The California Climate and Agriculture Network (CalCAN, 2017) is a coalition of sustainable agriculture organizations and farmer allies that come together to address the impacts of climate change on California agriculture and to find sustainable agriculture solutions. CalCAN is a network for farmers, providing information on everything from seeding and planting to irrigation, soil nutrient management, worker and labor safety, and financial planning. Water conservation, grazing management, and native grass restoration are a few of the topics discussed at CalCAN's annual Summit (CalCAN, 2017). California recently came out of a long drought, which exacerbated food insecurity, especially among communities of color. In April 2017 Californians passed the Farmer Equity Act of 2017, which recognizes that female farmers and farmers of color are socially disadvantaged and should be supported as other farmers are:

Existing federal agricultural policies have failed to provide sufficient and appropriate technical assistance and financial support, including farmer cooperative creation, for socially disadvantaged farmers and ranchers. It is therefore the intent of the Legislature that the secretary should support socially disadvantaged farmers and ranchers and include this support in the department's vision and its relevant policies (California Legislature, AB-1348 Farmer Equity Act of 2017, Section I, Article 6, 511 (a) 7 & (b), April 19, 2017).

The Farmer Equity Act of 2017 will help to cultivate climate resilience among vulnerable populations. It is a landmark piece of legislation in the EJ movement and should be a role model for other states.

CONCLUSION

To be climate resilient, a community must be able to adapt to and withstand climate change events such as a floods or hurricanes. Social cohesion provides networks that empower a community to work together to become climate resilient. New technologies enable cities to elevate roadways and divert water through culverts, as we saw in our case example of flood planning for Boulder, Colorado. A larger-scale implementation of technologies to create climate resilience is underway in the greater New York City metropolitan area devastated by Hurricane Irene and Superstorm Sandy.

Health equity is an integral part of resilience. To be resilient, communities need resources and access to medical and emergency response during and after a disaster. They need adequate access to water and food to be able to sustain themselves through a drought or flood. The United Nation's Sustainable Development Goals for 2030 years emphasize the connection between health equity and sustainability. A person's—and community's—health status is a function of socioeconomic status, and living and working conditions. As we saw in Chapter 4, health equity is a prerequisite for environmental justice, under which all people are afforded equal environmental protection.

REFERENCES

Arias, E. (2014). United States life tables, 2009. *National Vital Statistics Reports*, *62*(7), 1–63. PMID:24393483

Atkin, E. (2014, September 12). In Washington State, the 2014 wildfire season has been 6 times worse than normal. *ThinkProgress*, 2014. Retrieved from http://thinkprogress.org/climate/2014/09/12/3566740/washington-wildfires-worse-than-normal

Bachelet Jeria, M., & Halonen, T. (2016). *Breaking the ceiling for a better future*. Geneva, Switzerland: WHO.

Barron, J., Lipton, E., & Rivera, R. (2012, November 5). With $200 million in housing aid, officials begin relocating the displaced. *The New York Times*.

Bausson, D. (2015). *Social cohesion: the secret weapon in the fight for equitable climate resilience*. Washington, D.C.: Center for American Progress.

Brodt, S., Six, J., Feenstra, G., Ingels, C., & Campbell, D. (2011). Sustainable agriculture. *Nature Education Knowledge*, *3*(10), 1.

California Climate and Agricultural Network (CalCAN). (2017). Retrieved from http://calclimateag.org/

California Legislature. (2017). AB-1348 Farmer Equity Act of 2017, Section I, Article 6, 511 (a) 7 & (b), April 19, 2017. Sacramento, CA: CA State Legislature.

Carey, J. (2011). Storm warnings: Extreme weather is a product of climate change. *Scientific American*, *28*(June). Retrieved from http://www.scientificamerican.com/article/extreme-weather-caused-by-climate-change/

Denney, J. T., Kimbro, R. T., Heck, K., & Cubbins, C. (2017). Social cohesion and food insecurity: Insights from the Geographic Research on Wellbeing (GROW) Study. *Maternal and Child Health Journal, 21*(2), 343–350. doi:10.1007/s10995-016-2119-5 PMID:27439421

Food Assets for Assistance (FFA). (2017). World Food Program. Rome, Italy: World Food Program. Retrieved from http://www.wfp.org/food-assets

Freerice. (2007-2017). World Food Program. Rome, Italy: World Food Program. Retrieved from http://freerice.com/about/faq

Healthy People 2020. (2017). Social determinants of health. Washington, D.C.: U.S. Department of Health and Human Resources. Retrieved from https://www.healthypeople.gov/2020/about/History-and-Development-of-Healthy-People

Hine, A. C., Chambers, D. P., Clayton, T. D., Hafen, M. R., & Mitchum, G. T. (2016). *Sea Level Rise in Florida*. Gainesville, FL: University Press of Florida.

Ki-Moon, B. (2015). Foreword. In *Millennium Development Goals Report 2015*. New York: United Nations.

Klinenberg, E. (2002). Dying alone: an interview with Eric Klinenberg. Chicago: University of Chicago Press; Retrieved from http://www.press.uchicago.edu/Misc/Chicago/443213in.html

Klinenberg, E. (2013). Adaptation: How can cities be "climate-proofed?" The New Yorker, January 7, 2013. New York: Conde Nast. Retrieved from http://www.newyorker.com/magazine/2013/01/07/adaptation-2

Mann, H. (1843). Sixth annual report of the secretary of the board of education. *The Common School Journal, 5*(21), 193–352.

McAfee, T., Davis, K. C., Alexander, R. L., Pechacek, T. F., & Bunnel, R. (2013). Effect of the first federally funded anti-smoking national media campaign. *Lancet, 382*(9909), 2003–2011. doi:10.1016/S0140-6736(13)61686-4 PMID:24029166

Miniño, A. M., Murphy, S. L., Xu, J., & Kochanek, K. D. (2011). Deaths: Final data for 2008. *National Vital Statistics Reports, 59*(10), 1–126. PMID:22808755

Muskoka Initiative. (2014). Canada's ongoing leadership to improve the health of mothers, newborns, and children (2015-2020). Ottawa, Canada: Government of Canada. Retrieved from http://mnch.international.gc.ca/en/topics/leadership-muskoka_initiative.html

National Centers for Environmental Information. (2016a). Assessing the global climate in November 2016. Silver Spring, MD: US Department of Commerce. Retrieved from https://www.ncdc.noaa.gov/news/assessing-global-climate-november-2016

National Geographic. (2017). Sustainable Agriculture. Washington, D.C.: National Geographic. Retrieved from http://environment.nationalgeographic.com/environment/habitats/sustainable-agriculture/

NYRCRP. (2016). Round I & II. New York: Governor's Office of Storm Recovery. Retrieved from https://stormrecovery.ny.gov/nyrcr/final-plans

OECD. (2011). Perspectives on Global Development 2012 Social Cohesion in a Shifting World, Executive Summary. Paris, France: OECD. Retrieved from https://www.oecd.org/site/devpgd2012/49067839.pdf

Rael, A. (2013, September 23). Colorado flood damage: property loss estimated at around $2 billion. *The Huffington Post*. Retrieved from http://www.hufingtonpost.com/2013/09/23/colorado-flood-damage_n_3976222.html

Sastry, N., & Gregory, J. (2014). The location of displaced New Orleans residents in the year after Hurricane Katrina. *Demography*, *51*(3), 753–775. doi:10.1007/s13524-014-0284-y PMID:24599750

Schiavo, R., May Leung, M., & Brown, M. (2014). Community risk and promoting disease mitigation measures in epidemics and emerging disease settings. *Path Globe Health*, *108*(2), 76–94. doi:10.1179/2047773214Y.0000000127 PMID:24649867

South Florida Regional Planning Council. (2013). Adaptation Action Areas: Policy options for adaptive planning for rising sea levels. Southeast Florida Climate Compact. Retrieved from http://www.southeastfloridaclimatecompact.org/wp-content/uploads/2014/09/final-report-aaa.pdf

Sustainable Agriculture Research & Education (SARE). (2017). *What is Sustainable Agriculture?* College Park, MD: SARE.

The Tennesseean (2011). May 2010 flood:by the numbers. May 1, 2011. Retrieved from http://archive.tennessean.com/article/20110501/NEWS/305010091/May-2010-flood-by-numbers

United Nations. (2015a). We can end poverty: Millennium Development Goals and beyond 2015. Retrieved from http://www.un.org/millenniumgoals/news.shtml

United Nations. (2015b). Post-2015 Development Agenda. New York: United Nations. Retrieved from http://www.un.org/ga/search/view_doc.asp?symbol=A/69/L.85&Lang=E

Walker, B. & Salt, D. (2006). *Resilience thinking: sustaining ecosystems and people in a changing world.*

Whitcomb, D. (2014, September 18). New evacuations ordered as California wildfire doubles in size. Reuters. Retrieved from http://www.reuters.com/article/2014/09/18/us-usa-wildfires-idUSKBN-0HC21P20140918

White, A. (2013). Galveston still healing 5 years after Hurricane Ike. Galveston, TX: The Texas Tribune. Retrieved from https://www.texastribune.org/2013/04/26/galveston-still-healing-5-years-after-hurricane-ik/

World Health Organization. (2006). The World Health Report: Working together for health. Retrieved from http://www.who.int/whr/2006/en

World Health Organization. (2010). *Global Atlas of the Workforce*. Geneva, Switzerland: WHO.

World Health Organization. (2016a). Vector-borne diseases. Retrieved from http://www.who.int/mediacentre/factsheets/fs387/en/

World Health Organization. (2016b). Fact sheet: Malaria. Retrieved from http://www.who.int/malaria/media/world-malaria-report-2016/en/

World Health Organization. (2017). Global strategy for women's, children's and adolescents' health, 2016-2030. Retrieved from http://www.who.int/life-course/partners/global-strategy/global-strategy-2016-2030/en/

World Health Organization, Commission on Social Determinants of Health. (2008). Closing The gap in a generation; health equity through action on the social determinants of health. Executive Summary. Geneva, Switzerland: World Health Organization. Retrieved from http://apps.who.int/iris/bitstream/10665/69832/1/WHO_IER_CSDH_08.1_eng.pdf

KEY TERMS AND DEFINITIONS

Health Equity: The attainment of the highest level of health for all people.

Heat Index: An index that combines air temperature and relative humidity to calculate the "felt heat" impact on living organisms.

Heat Wave: A prolonged period of abnormally hot weather.

Infectious Disease: Illness that can be transmitted from one organism to another.

Millennium Declaration: In September 2000, world leaders adopted the UN Millennium Declaration, committing their nations to a new global partnership to reduce extreme poverty and achieve Millennium Development Goals by 2015.

Resilience: The capacity of a system to absorb disturbance and still retain its basic function and structure.

Social Cohesion: The set of characteristics that keep a group able to function as a unit to work towards the well-being of all its members, creating a sense of belonging and trust, and offering its members the opportunity of upward social mobility.

Social Determinants of Health: Create social and physical environments that promote good health for all; *Healthy People 2020* identifies five major determinants: 1) economic stability, 2) education, 3) neighborhood and built environment, 4) social and community context, and 5) health and health care.

Vector: An organism, typically a biting insect or tick, that transmits a disease or parasite from one animal or plant to another.

Vector-Borne Disease: An illness that is transmitted by a vector (e.g., malaria is transmitted by a mosquito, Lyme disease, by a tick).

Chapter 8
Climate Change Solutions:
Where Do We Go From Here?

ABSTRACT

Carbon pricing initiatives, as well as carbon capture and geologic sequestration (CCS), are tools to offset and reduce the impact of CO_2 emissions. The best solution is to not create the CO_2 emissions in the first place by switching from fossil fuels to renewable clean energy sources. This can be incentivized through tax breaks, as Norway has done with EVs. DOI can be used to change the public mindset so that they will embrace EVs, as Germany is doing now. Sea level rise solutions include shoreline armoring and beach renourishment, elevation of roadways and sidewalks, managed retreat through purchase of vulnerable land for public use, and avoidance through limiting development in high-risk areas. This chapter gives case examples from U.S.'s 100 Resilient Cities, and UK's Bristol is Open, a programmable city where data on air quality, transportation, health, and needs of elderly residents are integrated into one high-speed centralized network.

INTRODUCTION

Short of building an ark, where do we go from here? The consensus among most experts is that it is too late to stop climate change. A twinned approach to climate change is recommended, one that involves both adaptation and mitigation. As the late Australian epidemiologist Anthony McMichael (2010)

DOI: 10.4018/978-1-5225-3414-3.ch008

put it, "Adaptation is managing the unavoidable; mitigation is avoiding the unmanageable."

Although the situation is dire and there is a high probability that we can only slow down global warming, not stop it, there are some glimmers of hope. Returning to the framework of DOI Theory, we must thank the tenacious persistence of the innovators and early adopters (DOI Theory—see Chapter 5), who "got it" about climate change back in the 1970s and have continued to pursue regulations to limit GHG emissions for the past forty years or so. The United States had a wakeup call during the 1973 oil embargo by OPEC, which caused gasoline shortages, long lines at the pump, and much higher gas prices. At the height of the energy crisis, everyone had to conserve gasoline, as supplies were limited. Once the crisis eased, most people resumed their usual consumption habits. However, the early adapters became more ecologically conscious, starting organic food coops, composting leftover food scraps, recycling, biking instead of driving, installing solar panels, and urging the public to reduce consumption of fossil fuels. Automobile manufacturers, however, catered to the late adopters, who were in the majority, by producing bigger SUVs, which guzzled more gas and emitted more CO_2.

Environmentalist and author Bill McKibben was one of the first to write about global warming for the general public; his 1989 book, *The End of Nature*, warned the world that their "reassuring sense of a timeless future" is a delusion, due to rapidly rising CO_2 levels in the atmosphere, a hole in the ozone, and acid rain. In 1992, Senator Al Gore's book, *Earth in the Balance,* warned about the dangers of deforestation in the Amazon and of "needles, dead dolphins, and oil-soaked birds" on beaches, evidence of the environmental degradation of our waterways. As Vice President from 1993 to 2001, Gore helped to push forward protective environmental legislation, including President Clinton's 1994 Executive Order 12898 on Environmental Justice (see Chapter 4 herein), and the 1996 Mercury-Containing and Rechargeable Battery Management Act (P.L. 104-19, Food Quality Protection Act (amended FIFRA), and *Safe Drinking Water Act Amendments.*

As a result of growing worldwide concern over global warming, the United Nations Framework Convention on Climate Change (UNFCCC) was established in 1994 with the aim of "stabilization of greenhouse gas concentrations in the atmosphere at a level that would prevent dangerous anthropogenic [human-induced] interference with the climate system" (United Nations, 2014a). The first binding UNFCCC agreement to reduce

GHG emissions was the 1997 Kyoto Protocol, which went into effect in 2005, with its first commitment period from 2008 to 2012. It mandated reduced emissions by 37 industrialized nations and the European Community; the United States and Australia did not ratify it because of the lack of regulation of emissions from developing countries.

In 2006, Gore's book and documentary film, *An Inconvenient Truth: The Planetary Emergency of Global Warming and What We Can Do About It,* reached hundreds of millions of people around the globe, greatly increasing world consciousness about the imminent dangers of global warming. In 2008, Dr. James Hansen of NASA, widely regarded as the most eminent climate scientist in the United States, wrote an article with his colleagues entitled, *Target Atmospheric CO2: Where Should Humanity Aim?* Based on analysis of the scientific evidence on the impact of CO_2 in the atmosphere, Hansen and his colleagues concluded that

If humanity wishes to preserve a planet similar to that on which civilization developed and to which life on Earth is adapted, paleoclimate evidence and ongoing climate change suggest that CO2 will need to be reduced from its current 385 ppm to at most 350 ppm.

Since 2008, when Hansen and his colleagues wrote this article, the CO_2 concentrations in the atmosphere observed at Mauna Loa have continued to rise, from 385 ppm to monthly averages of 404.04 ppm in February 2016 and 406.42 ppm in February 2017 (NOAA, 2017). Recent average monthly trends are shown in Figure 1.

The CO_2 concentrations are now well above the limit that Hansen and his colleagues deemed to be calamitous for the planet.

The nonprofit environmental activist organization, 350.org, founded by Bill McKibben in 2008, is raising world awareness about the urgent need to reduce CO_2 levels. In 2012, McKibben wrote an influential article in *Rolling Stone* (2012) to draw attention to the "dangerous math" of global warming. In 2013, his movie, *Do the Math,* helped to raise world awareness about the global warming crisis precipitated by increasing CO_2 concentration in the atmosphere. McKibben (2013) stated that the goals of 350.org are twofold: 1) to freeze new fossil fuel investments, and 2) to wind down existing fossil fuel investments in five years. McKibben's organization started at Middlebury College, Vermont and has been popular on college campuses; 350.org gets its message across through civil disobedience and mass demonstrations, as well as global art installments in which thousands of people form human

Figure 1. Monthly average CO_2 concentration in the atmosphere at Mauna Loa
Source: National Oceanic Atmospheric Administration (NOAA, 2017).

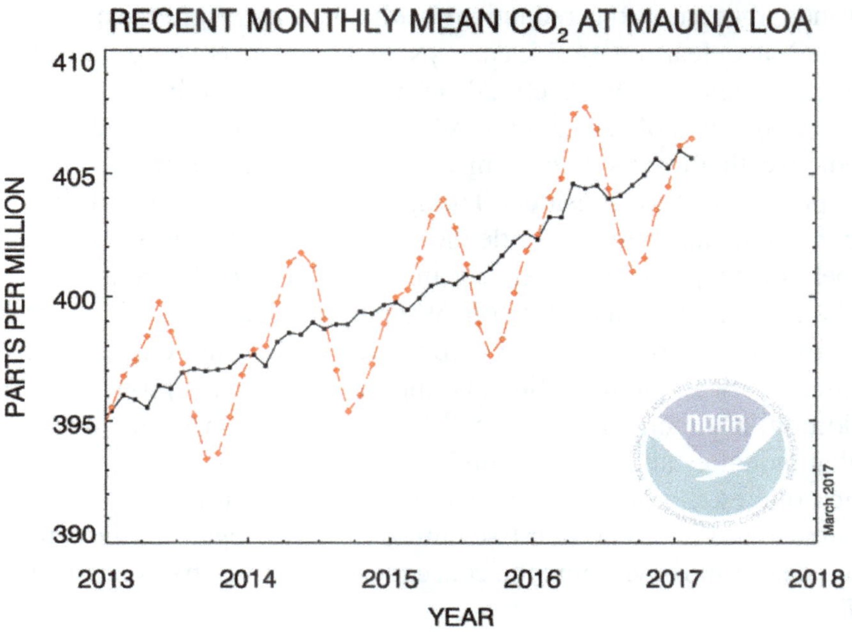

signs that can be seem from outer space with messages about ending fossil fuel dependence (McKibben, 2013).

A TIPPING POINT

After years of stalemates at global climate conferences, the nations of the world gathered at the historic UNFCCC in Paris in 2015 and signed the Paris Agreement. Adopted on December 12, 2015, it signaled a tipping point in the climate change movement, with a worldwide pledge to reduce GHG emissions enough to limit global warming to a maximum of 2 degrees Celsius (3.6 degrees Fahrenheit) (United Nations, 2015). The parties to the Paris Agreement submitted Intended Nationally Determined Contributions (INDCs) that quantified each nation's target reduction in GHG emissions, and in November 2016, the Paris Agreement went into effect. The Marrakesh Partnership for Global Climate Action was established to help catalyze climate action by Parties and non-Party stakeholders in 2017-2020 (United Nations, 2016). The challenge now is, how are the nations of the world to meet the target emissions reductions set by the Paris Agreement? The keys are to embrace clean, renewable energy sources, while reducing waste through increased efficiency (Maibach, 2017). It is also imperative to adapt to the inevitable impacts of rising sea levels, rising temperatures, and increasing extreme weather events, to minimize loss of life and property.

At presstime, U.S. President Trump was threatening to pull out of the Paris Agreement and to dismantle the Clean Power Plan. Based on the huge numbers of people who turned out in April 2017 for the Science March, and the People's Climate March in Washington, D.C. and 370 other cities nationwide in support of initiatives to reduce GHG emissions, one can infer that the American public is fiercely opposed to Trump's proposed actions. In addition, members of the House Climate Solutions Caucus introduced a bill that would establish a 10-member commission to conduct a review of public policies and private actions to reduce GHG emissions and to make recommendations on the latest scientific findings regarding how to avoid serious health and environmental consequences from global warming (Hand, 2017).

There is some good news: after decades of increasing global CO_2 emissions, there was a flattening trend in global CO_2 emissions in 2015-2016, due to

Figure 2. Trends in global carbon emissions 1980-2016
Source: International Energy Agency (IEA, 2017).

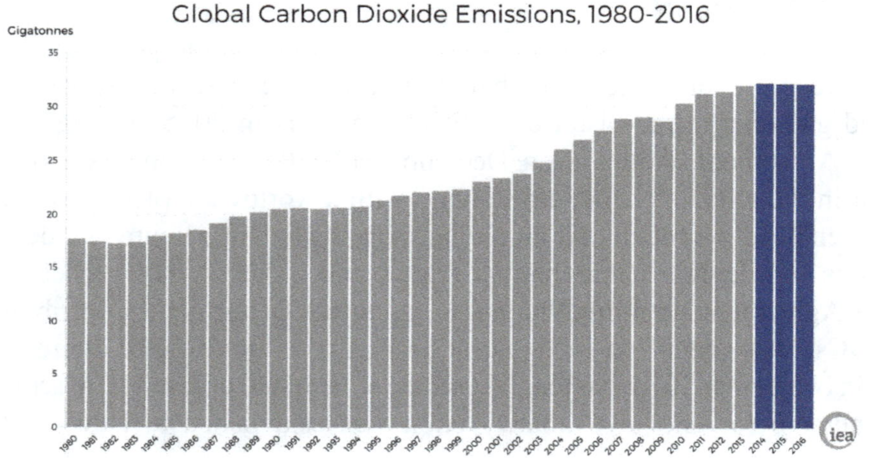

worldwide investment in renewable, sustainable energy sources (IEA 2017). Scientists consider CO_2 to be the principal determinant of Earth's climate state, the "control knob" that sets global mean temperature (Lacis et al., 2010, 2013). Figure 2 shows trends in global carbon emissions from 1980 to 2016.

The world has begun to shift its energy consumption; in 2016, renewable energy sources supplied more than 50 percent of the global electricity demand, with half of that coming from hydropower. Coal demand fell worldwide, especially in the United States, where it was down 11 percent in 2016 (IEA, 2017). Nevertheless, the CO_2 levels remain beyond what leading scientists believe is sustainable for life as we know it on earth. More needs to be done! One solution involves putting a price on carbon.

PUTTING A PRICE ON CARBON

Carbon pricing means factoring in the actual costs of CO_2 emitted in manufacturing products; without carbon pricing, pollution is essentially free. The strategy behind carbon pricing is to price products according to their real cost; the result is that, with carbon costs added into the price, "dirty" products created by burning fossil fuels will be more expensive than "clean" ones created with renewable energy, and businesses will be motivated to find new ways to reduce pollution. Initially, when carbon pricing is put into place, there have to be tax cuts and increased payments to households to help pay for the increased cost of living.

The World Bank (2015, p. 17) wrote the following about carbon pricing:

There is a growing consensus among both governments and businesses on the fundamental role of carbon pricing in the transition to a decarbonized economy. Placing an adequate price on GHG emissions helps mobilize the financial investments required to support diverse actions, such as fuel switching from coal to natural gas, renewable energy deployment, the adoption of energy efficiency measures and the use of low-carbon technologies in industry.

Carbon pricing initiatives include emissions trading systems (ETSs), carbon taxes, carbon offsetting with carbon credits, results based finance (RBF), and internal carbon prices set by companies. China and the United States have the largest volume of emissions covered by carbon pricing instruments; in China, carbon pricing instruments cover 1 gigaton (Gt) CO2 equivalent (e), while in the United States they cover 0.5 GtCO2 e (World Bank, 2015).

China is moving to a national emissions trading system (ETS); it currently has seven pilot projects, which together form the largest national carbon pricing initiative in terms of volume. An ETS (also known as a cap-and-trade system) caps the total level of greenhouse gas emissions and allows those industries with low emissions to sell their extra allowances to larger emitters (World Bank, 2015). The European Union Emissions Trading System (EU ETS), which covers 2 GtCO2 e of emissions, is the single largest international carbon pricing instrument (World Bank, 2015). In 2015, 29 national and 23 subnational jurisdictions around the world were using carbon pricing; the total volume covered, 7 GtCO2 e, represented a 90 percent increase since 2012. However, this total is still only about 12 percent of annual global GHG emissions (World Bank, 2015).

Carbon Tax

A carbon tax is a fee on fossil fuel emissions; users of fossil fuels pay for the climate damage their fuel use imposes by releasing CO_2 emissions into the atmosphere. The intention behind such a tax is to motivate businesses to switch to clean energy (Carbon Tax Center, 2017). Climate scientist James Hansen has repeatedly made the case for a global tax on carbon, stating that polluting should not be free. We pay to have a garbage truck remove our household waste. If we decided to throw our garbage in the street instead, we would be fined. Yet oil and coal companies are polluting our atmosphere for free. Hansen said, “We are allowing fossil fuel companies to use the atmosphere as a free waste dump.” Fisher Stevens, director of the 2016 global warming documentary *Before the Flood*, starring Leonardo DiCaprio, applauds the use of carbon taxes to move the world in the direction of clean energy: “[Y]ou can write your senator or congressman and say, look, we believe in a carbon tax. Because that financial disincentive for using fossil fuels is one of the main things that can help the U.S. move to the next generation of energy” (Stevens, 2016).

In Europe, Sweden, Norway, The Netherlands, Denmark, Finland, Austria, Germany and Italy, Slovenia, and the UK all have carbon taxes. India, Japan, Costa Rica, and Zimbabwe all have imposed carbon taxes. Chile has a carbon tax but it has not yet taken effect. Australia began a carbon tax program in 2012, but repealed it in 2014. The United States does not have a carbon tax.

Carbon Offsets

Carbon offsets are bought and sold by carbon offset providers, international brokers, and retailers, to counteract CO_2 emissions released by the burning of fossil fuels. Since the Kyoto Protocol went into effect in 2005, parties to it have to reduce CO_2 emissions and one way to do this is to buy carbon offsets, which neutralize the users' CO_2 emissions. It is important to check the credentials of carbon offsetting companies, as not all of them are reliable. Some disreputable carbon offset companies sell carbon credits from non-existing carbon reduction projects, fund renewable energy projects that would have happened anyway, or sell the same credits several times over (Allianz, 2017). One way to verify quality is to select a company that has been awarded the Gold Standard, an international standard for carbon emission reductions that ensures that offset investments help the environment. However, Gold Standard projects can only take place in countries that do not have to reduce emissions under the Kyoto Protocol (Allianz, 2017).

Some seek to reduce carbon offsets by planting forests. However, forest fires often cancel out the benefits of plantings, and some of the areas most in need of carbon offset are already plagued by drought and would have to divert water from elsewhere to keep the forests healthy. One such example is China's March 23, 2017 announcement that it plans to create a "green necklace" of trees in the Hebei Province surrounding Beijing to help clear the air of smog (Wong, 2017). Hebei Province has coal-powered steel factories that have made it home to the most polluted cities in China. The emissions from the steel factories have caused severe smog in Beijing, a city of 22 million, and other cities in northern China. The Hebei provincial government states that it will tap into rivers, reservoirs, wetlands and farmland in order to create the new forest. It does not mention that Beijing suffers from a persistent drought, and that there is very little water available (Wong, 2017).

Given the problems associated with carbon offsetting, it is regarded by many as a second-best solution to reducing greenhouse gas emissions. The best solution is not to create the emissions in the first place. The harmful emissions that are released into the atmosphere have a lasting effect; no magician is going to make them disappear. In sum, offsetting should only be used when CO_2 emissions cannot be reduced (Allianz, 2017).

Do you know your carbon footprint (how much carbon you use in daily activities such as driving, heating your home, and eating meat (carbon-intensive food)? There are many online carbon footprint calculators, some tailored to specific age groups. Environmental groups such as the U.S. Environmental Protection Agency and the Nature Conservancy post carbon footprint calculators. Calculating your own carbon footprint shows how your household measures up compared to the average household in your zipcode; it also pinpoints which areas of your life add the most carbon to your footprint, so that you can develop strategies to reduce areas of high carbon usage. If you want to monitor your carbon footprint while on the go, a phone app is useful. Phone apps ask a series of questions about travel, commuting, eating and other habits, and then calculate an appropriate annual contribution to offset carbon usage.

Individual consumers also can purchase carbon offsets to neutralize their personal CO_2 emissions. For example, driving 2,000 miles (3218 km) in a mid-sized car releases about a ton of CO_2. Offsetting these emissions costs about five dollars. Some airlines offer passengers the option to neutralize airplane emissions by buying carbon offsets along with their tickets. For example, offsetting a return flight from the UK to New Zealand would cost about 116 dollars (Allianz, 2017). The money from the carbon offset goes to projects that produce renewable energy or reduce carbon emissions. Innovative technology has made calculating one's personal carbon offset a snap.

Carbon pricing is a tool that should be used in tandem with other policy instruments such as the removal of fossil fuel subsidies, renewable energy portfolio standards, and energy efficiency standards (World Bank, 2015). Another method of reducing CO_2 levels in the atmosphere is carbon capture and (geologic) sequestration.

CARBON CAPTURE AND (GEOLOGIC) SEQUESTRATION

Carbon capture and sequestration (CCS) is process that reduces the release of CO_2 emissions from industrial sources into the atmosphere by capturing the CO_2 from power plants or industrial processes, transporting it via pipeline, and then injecting it underground for geologic sequestration (storage) in deep underground rock formations a mile or so beneath the surface. Impermeable layers of rock above the sequestration site trap the CO_2 and prevent it from migrating upward. CCS can reduce CO_2 emissions from fossil fuel burning power plants by 80 to 90 percent (U.S. EPA, 2016a). Approximately one third of the United States' carbon emissions come from power plants and other large point sources. The U.S. Department of Energy's Carbon Storage Program is focused on ensuring safe and permanent storage and/or utilization of CO_2

captured from point sources through Cooperative Advancement, which seeks to advance CCS worldwide, and Carbon Storage and Utilization Research, which conducts R&D to find potential commercial reuses of stored CO_2. (U.S. Department of Energy, 2017).

To date, the costs of CCS have been high because of the initial expense incurred in building the capture and compression facilities. For example, the SaskPower's Integrated Carbon Capture and Storage facility in southeastern Saskatchewan cost some 1.35 billion Canadian dollars (Global CCS Institute, 2016). The Global CCS Institute in Australia has identified 38 large-scale CCS projects around the world; 20 of these are expected to be operational by the end of 2017. Two CCS projects launched in 2016 were the Abu Dhabi CCS Project, Phase 1, the world's first large-scale application of CCS to iron and steel production, and Japan's Tomakomai CCS Demonstration Project, with capture at a hydrogen production facility and near-shore storage (Global CCS Institute, 2016). Three more major large-scale CCS projects are poised to begin operations in the United States: 1) the largest post-combustion capture project at a power station (the Petra Nova Carbon Capture Project in Texas); 2) the first large-scale bio-CCS project (the Illinois Industrial Carbon Capture and Storage Project); and 3) the world's first CCS project at a commercial-scale coal gasification power facility (the Kemper County Energy Facility in Mississippi). Other large-scale projects are underway in Canada, Europe, South America, Australia and parts of Asia and the Middle East (Global CCS Institute, 2016),

The Intergovernmental Panel on Climate Change (IPCC) Fifth Assessment Report Summary for Policymakers found that CCS was a crucial part of the world's ability to meet the emissions reduction targets of the Paris Agreement and that, without CCS, mitigation efforts would cost double (IPCC, 2014). The International Energy Agency (IEA, 2013, p. 8) stated:

CCS is currently the only large-scale mitigation option available to make deep reductions in the emissions from industrial sectors such as cement, iron and steel, chemicals and refining. Today, these emissions represent one-fifth of total global CO2 emissions.

The IEA (2013) found that almost 4,000 million tonnes of CO_2 per year (Mtpa) need to be captured and stored by 2040 to meet a 2 degree Celsius scenario. The IEA recommended R&D to make CCS more efficient and economical.

While CCS is a useful and important part of CO_2 emissions reduction, the key to achieving the Paris Agreement target of 2 degrees Celsius temperature increase is to end reliance on fossil fuels and incentivize the switch to clean, renewable energy sources.

INCENTIVIZING THE SWITCH TO CLEAN ENERGY

To achieve the Paris Agreement goals, mitigation is needed; manufacturers and consumers need to be incentivized to switch from fossil fuel energy sources to clean, renewable energy sources. Incentives can take several forms, for example, discounts on purchases of EVs, or tax breaks for construction companies and homeowners who invest in solar energy and green building materials. In the past few years, EV sales have increased substantially in Norway, making this country the world's leader in EV usage. In contrast to standard internal combustion engines, which account for about 14 percent of human-induced global GHG emissions (IPCC, 2014), electric vehicles have zero emissions. Norway's EV incentive program can serve as a role model for the world, especially for industrialized nations such as the United States, where transport (e.g., cars, trucks, SUVs, minivans, trains, ships, airplanes) is responsible for 26 percent of GHG emissions (U.S. EPA, 2016b).

Embracing New Technology: Norway's Electric Vehicle (EV) Revolution

Norway is far and away the world leader in adoption of EVs, with a market penetration of 23.3 percent (IEA, 2016). Sales of EVs have been booming, thanks to a government incentive program that offers generous subsidies, including tax exemptions on EV purchases and free use of toll roads for EVs: "It has to be more expensive to pollute than to use environmentally friendly fuels" (Jolly, 2015). Norway's vehicle taxes are extremely high, including a sales tax of 25 percent plus a registration tax that averages $12,000 or more (Jolly, 2015). EVs are exempt from the high tolls for bridges and tunnels, too. In addition, gasoline costs more than the equivalent of $6 in Oslo, whereas charging an EV battery is inexpensive. The incentive program is costing the country revenues and will be phased out gradually beginning in 2018 (Mirani, 2015). The intention is to offer the incentive program long enough to change the way people think about transportation (Jolly, 2015).

Embracing Agricultural Innovations: Israel's Water Revolution

Israel is far and away the world leader in water management thanks to its embrace of wastewater reclamation and reuse, as well as its invention and widespread adoption of drip irrigation and other technological innovations. Israel's water revolution has turned around an arid, drought-stricken land that is 60 percent desert into a water management leader. The five principles of the Israeli water revolution can guide other regions facing water shortages. They are: 1) education and conservation, beginning in school and also on media, 2) wastewater reclamation and reuse (86 percent of wastewater is reused for agriculture), 3) centralized government control over water sources, 4) innovative, energy-efficient agricultural technology, and 5) desalination. Israel's water revolution success rests on its combination of education, conservation, and the use of innovative technology (Distel, 2016). Israel's water recycling provides a role model for other drought-stricken regions (Grossman, 2016). Spain, which is second in wastewater reclamation and reuse, recycles only 17 percent of its water, and the United States, only 1 percent of its water (Kershner, 2015).

PROMOTING URBAN RESILIENCE

Urban resilience was one of the themes at the February 17, 2017 Health & Climate Change Meeting in Atlanta, Georgia, organized by Former Vice President Al Gore. The world is becoming more urban every day; more than half of the world's population lives in cities and each week 1.5 million people move to cities (United Nations, 2014b). Cultivating urban resilience helps communities to adapt to climate change. Gore mentioned the terrible toll of the 1995 Chicago heat wave, which claimed more than 700 lives, mostly among poor and elderly residents without air conditioning (Klinenberg, 2002). In the aftermath of this calamitous experience, Chicago worked hard to cultivate urban resilience and experienced greatly reduced casualties in a subsequent severe heat wave (Klinenberg, 2013). Chicago is now one of the Rockefeller Foundation's *100 Resilient Cities* (100 Resilient Cities, 2017). The Rockefeller Foundation invested $100 million in *100 Resilient Cities* in 2013 to create urban resilience, which it defined as follows:

Urban resilience is the capacity of individual, communities, institutions, businesses, and systems within a city to survive, and grow no matter what kinds of chronic stresses and acute shocks they experience (100 Resilient Cities, 2017).

Chronic stresses that plague cities on a daily or cyclical basis include high unemployment, overtaxed or inefficient public transit, endemic violence, and chronic food and water shortages; acute stresses that arrive suddenly and threaten a city include earthquakes, floods, epidemics, and terrorist attacks (*100 Resilient Cities*, 2017).

Urban resilience has four dimensions: 1) health and well-being, 2) economy and society, 3) infrastructure and environment, and 4) leadership and strategy. The *100 Resilient Cities* program focuses on infrastructure and social cohesion. It uses the Building Resilience Against Climate Effects (BRACE) Framework of the Centers for Disease Control and Prevention (CDC), which is shown in Figure 3.

Figure 3. CDC's BRACE framework
Source: Centers for Disease Control and Prevention (2015).

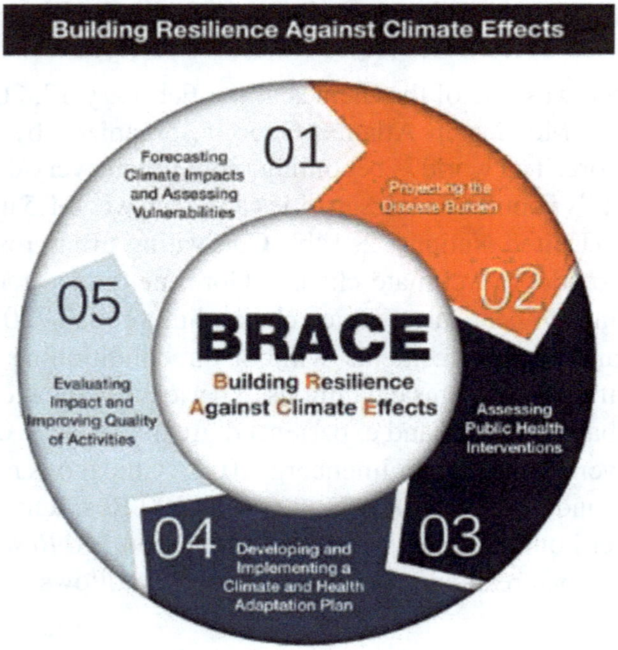

BRACE is a five-step model: Step 1 is to anticipate climate impacts and assess vulnerabilities; Step 2 is to estimate the additional disease burden associated with climate change; Step 3 is to identify the most suitable health interventions for the identified health impacts; Step 4 is to develop, disseminate, and implement a written climate action plan, and update it regularly, and Step 5 is to evaluate the impact and improve the quality of activities (CDC, 2015).

The BRACE Framework is tailored to each city, to address the specific climate resilience needs of its population and infrastructure. To foster urban resilience, the city's individual systems need to be strengthened to be able to withstand, respond to, and adapt readily to stresses and shocks. Resilient systems share seven qualities: they are reflective (using past experience to inform future decisions), resourceful (identifying alternative resources), inclusive (utilizing shared decision making), integrated, robust, redundant (with backup capacity to accommodate disruption), and flexible (*100 Resilient Cities*, 2017).

Following are two case examples from the *100 Resilient Cities* initiative that demonstrate how urban resilience to climate change is being fostered in the southeastern United States and southwest United Kingdom.

Case Example 1: Adapting to Rising Sea Levels in Greater Miami and the Beaches

The major challenges facing this Southeastern Florida coastal city and its beaches are impacts from rising sea levels, coastal erosion, and hurricanes. Greater Miami and the Beaches (GM&B) comprise Miami-Dade County, the City of Miami, and the City of Miami Beach. While sea level rise is an issue facing all coastal communities, the State of Florida presents a special case because it

- Is the flattest of the contiguous United States, with its highest point 106 meters above sea level (Dobson & Campbell, 2014);
- Is the contiguous state with by far the most tidal shoreline;
- Is buffeted by hurricanes and tropical storms from the Atlantic Ocean to the east and the Gulf of Mexico to the west;
- Is underpinned with porous limestone rock;
- Has a shallow water table whose depth and salinity strongly impact where different animals and plants can live;
- Is home to a rich biodiversity of rare plants and animals;

- Has a rapidly increasing population, much of in coastal communities (Clayton, 2016).

The major impacts of sea level rise in GM&B are flooding, especially during storms and biannual King Tides, saltwater intrusion into freshwater sources, water table elevation, and beach erosion; if sea level rise continues without mitigation, much of the County will be inundated (permanently converted to wetlands or open water) (Clayton, 2016).

Miami Beach, a major tourist destination, recently invested some $400 million into elevating sidewalks and installing pumps to keep roadways and pedestrian walkways dry during high tides and storms. Miami Beach Mayor Philip Levine was interviewed by Leonardo DiCaprio in *Before the Flood* (2016), an environmental documentary about global warming; Mayor Levine said that the Miami Beach pumps project will last for 40 years or so with the current rate of sea level rise; it is an adaptation to current conditions. He hopes that, in the meantime, a more permanent solution will be found (*Before the Flood*, 2016).

The *100 Resilient Cities* initiative helps each city to develop a Resilience Strategy Plan. In GM&B three Chief Resilience Officers--Jane Gilbert for the City of Miami, Susanne Torriente for the City of Miami Beach, and James F. Murley for Miami-Dade County--will develop GM&B's first comprehensive Resilience Strategy Plan to prepare GM&B to handle "shocks" (catastrophes such as hurricanes, fires, and floods) and "stresses" (ongoing issues such as affordable housing, homelessness, and unemployment).

In February 2017, it was announced that Miami was one of five U.S. cities awarded a Smart Cities Readiness Challenge Grant (Peeples, 2017). To help deal with tidal flooding, Miami will use the Smart Cities grant to implement a high-tech Sea Level Rise Pilot Program in collaboration with the Environmental Systems Research Institute (ESRI). The program will integrate Geographic Information Systems (GIS), 3-D modeling, waterfront sensors, and LIDAR (light detection and ranging) data to identify and map out the most vulnerable areas and provide residents with timely flooding alerts (Peeples, 2017). In LIDAR, an airborne laser is pointed at a target area on the ground and a beam of light is reflected by the surface it encounters. The data collected are used to create a "point cloud" of elevation points, which, in turn, is used to generate geospatial models. LIDAR is useful for mapping shorelines and for emergency response (NOAA, 2015).

Case Example 2: Using New Technologies for Education and Social Justice in Bristol, UK

Bristol is a mid-size city in southwest England, and the only UK city to win the European Green Capital Award (2015). Bristol Mayor Marvin Reeves notes that Bristol has consistently been voted one of the best places to live. However, there are pockets of poverty: 42 areas in Bristol are among the 10 percent most deprived in England, and 16 percent of Bristol's residents live in poverty. There is a persistent life expectancy gap between the most and least deprived areas of Bristol (approximately 9.6 years for men and 7 years for women). Mayor Reeves states, "My vision is to build a livable city of opportunity built on fairness and inclusion" (Bristol Resilience Strategy, 2016, Forward). Education is a key component of social justice and socioeconomic upward mobility. In 2016, Bristol became the first city in England to become part of the UNESCO Global Network of Learning Cities (Bristol Resilience Strategy, 2016). UNESCO's Global Network of Learning Cities was established after the International Conference on Learning Cities in Beijing, China adopted the Beijing Declaration on Building Learning Cities and issued a Call to Action to UNESCO to form a global coalition and to the nations of the world to join it (UNESCO, 2013). Subtitled, "Lifelong Learning for All: Promoting Inclusion, Prosperity and Sustainability in Cities," it includes 12 commitments, which "have the power to transform our cities" (UNESCO, 2015):

- Fostering social cohesion through universal literacy, gender equality, civic engagement, and the creation of a safe, harmonious and inclusive community.
- Enhancing economic development and cultural prosperity.
- Promoting sustainable development to ensure the future viability of communities.
- Promoting inclusive learning in the education system.
- Revitalizing lifelong learning in families and communities.
- Facilitating learning for and in the workplace to keep up with technological advances.
- Extending use of information and communication technologies (ICT) such as the Internet.

- Providing learner-friendly environments and supporting learners with special needs.
- Fostering lifelong learning in museums, libraries, parks, and cultural centers.
- Strengthening political will and commitment.
- Improving governance and participation of all stakeholders.
- Investing in lifelong learning for its wider benefits on public health, economic growth, reduced criminality, and increased democratic participation.

Bristol's resiliency framework projects 50-year targets for 2066. In 2066, Bristolians will all have affordable housing, job opportunities, health equality and be free from discriminatory practices (Bristol Resilience Strategy, 2016). The city will be carbon neutral, with zero waste, have a green infrastructure, clean air, and will use sustainable supply chains and transport. In the near-term, Bristol will address its homeless problem by increasing capacity at all levels. It will foster cultural engagement and social cohesion through "Migrant Dialogues," an art-led initiative by a partner NGO to build a dialogue about immigration, diversity and neighborliness. It will give 16 and 17 year olds the right to vote to empower and engage young people (Bristol Resilience Strategy, 2016).

A technological innovation, REPLICATE (REnaissance of Places with Innovative Citizenship And Technologies), a Smart city 'Lighthouse' project, will use digital technology to explore the impact of integrating smart energy and smart transport interventions in the impoverished neighborhood of Easton. Bristol is part of the URBACT Resilient Europe network, whose members work together, sharing best practices, to achieve resilience and sustainability. Bristol will provide free bus transportation for youth under age 16, and will employ community-based adaptation (CBA) techniques to build capacity in marginalized communities. All major initiatives and investments are subject to an Environmental Impact Assessment (EIA) and an Equalities Impact Assessment (EQIA) (Bristol Resilience Strategy, 2016).

Another technological innovation is *Bristol is Open Network,* an R&D digital test pad to develop and test plans for a programmable city, where data on air quality, transportation, health, need of elderly residents, will be integrated into one high-speed centralized network (Bristol Resilience Strategy, 2016).

SEA LEVEL RISE SOLUTIONS

Global mean sea level rose by 19.5 cm (7.7 inches) from 1901 to 2015, at an average rate of 1.7 mm (.07 inches)/year, but with significant decadal variation. The rate of sea level rise since 1993, when satellite measurements became available, has been higher, at around 3 mm (.12 inches)/year. Global mean sea level in 2015 was the highest yearly average over the record and ~70 mm (2.8 inches) higher than in 1993 (European Environment Agency, 2017).

There is a high correlation between trends in sea level rise, mean global temperature, and CO_2 concentration in the atmosphere. Figure 4 shows these historic trends from 400,000 years ago to the present time.

Sea level rise is not uniform all over the world due to land shifts, but it is inevitably viewed by people who live in the interior as "a coastal problem" and not something that affects them directly. For this reason, it is vital for coastal communities to join forces to create adaptation plans and climate resilience. One such network is the Southeast Florida Regional Climate Change Compact (The Compact), a partnership of four southeastern Florida counties—Broward, Miami-Dade, Monroe, and Palm Beach—formed to work collaboratively on mitigation and adaptation strategies. The Compact

Figure 4. Historic trends of global temperature, CO_2 in the atmosphere, and sea level
Source: JohnEnglander.net (2017).

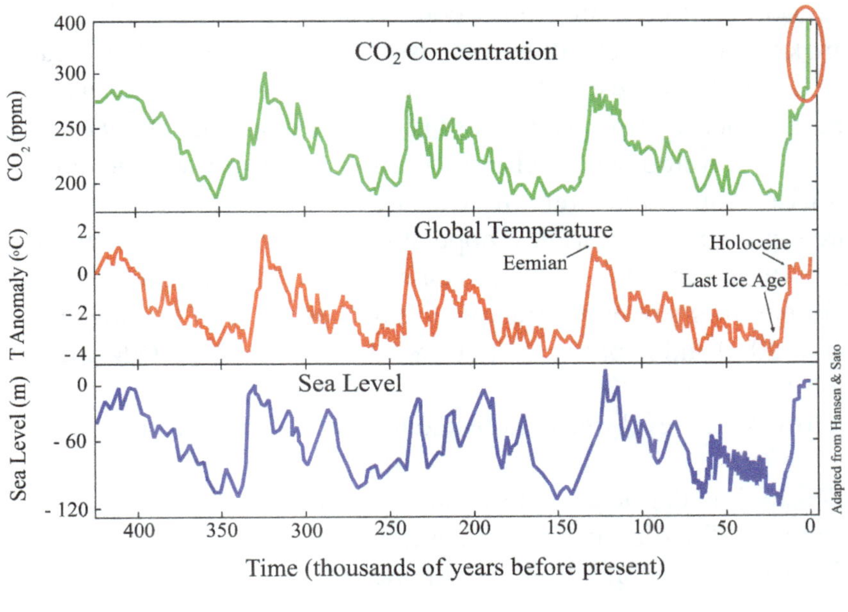

has been working on changing Florida planning legislation to incorporate efforts to address climate change and sea level rise, and to prioritize funding for these efforts (South Florida Regional Planning Council, 2013). In 2011, the Florida Legislature passed the Community Planning Act (CPA), which addressed coastal hazards planning, with designated Adaptation Action Areas prioritized for funding. However, climate change progress has been blocked by Governor Scott, who denies the reality of human-induced global warming. Much more needs to be done in Florida to prepare for the impacts of sea level rise, as it has the most coastline of any state and is the flattest contiguous US state (South Florida Regional Planning Council, 2013).

Community-based NGOs have taken up the slack. For example, the Unitarian Universalist Fellowship is serving as the umbrella organization for *Rising Together*, an environmental justice project of the U.S. Environmental Protection Agency (http://www.uufbr.org/green-sanctuary-team/). Using a sea level rise survey developed by Dr. Keren Prize Bolter of Florida Atlantic University, in 2016 canvassers surveyed respondents in vulnerable coastal communities in southeast Florida about their perceived risk from sea level rise and their experiences with flooding, water contamination, mold, and asthma. *Rising Together* partnered with Habitat for Humanity, which helped lower-income residents whose homes had sustained storm damage to repair and rebuild. *Rising Together* is being replicated in 2017 in other vulnerable communities in central and south Florida; a toolkit has been developed to empower communities to address issues associated with sea level rise (www.reacttoolkit.net). In Hollywood, Florida, where the biannual king tides make coastal streets all but impassable, Temple Solel has created the *Higher Ground Initiative* to work with the *Sea Level Rise Solutions Project* to bring together Jewish synagogues in coastal Florida and along the Eastern Seaboard to develop sea level rise solutions (https://www.facebook.com/sealevelrisesolutions/?hc_ref=SEARCH).

Three-quarters of Florida's population lives in coastal counties that generate 79 percent of the state's total revenues. The replacement value of the built environment and infrastructure in these coastal counties was $2.0 trillion in 2010 and is projected to be $3.0 trillion in 2030 (Florida Oceans and Coastal Council, 2010).

In Florida, sea level rise is expected to have four major impacts:

- Coastal inundation and shoreline recession.
- Increased flooding from severe weather events.
- Saltwater intrusion into water supplies.

- Elevated coastal groundwater tables.

Adaptation planning includes protection strategies such as shoreline armoring and beach renourishment, accommodation such as the elevation of certain roadways and sidewalks in Miami Beach, managed retreat through purchase of vulnerable land for public use, and avoidance through limiting development in high-risk areas. Building has to adhere to floodplain regulations, with setbacks (certain distances required from shoreline feature) and buffers (undeveloped land such as wetlands that supports natural functions). Incentives such as tax credits and Transfer of Development Right compensate owners for giving up the right to develop part or all of a property (South Florida Regional Planning Council, 2013).

The beach erosion in coastal Florida is exacerbated by the removal of natural protective factors in the marine ecosystems. One hundred years ago, mangrove forests lined the shores of both the Atlantic coastline and the Gulf Coast of Florida (Florida Department of Environmental Protection, 2017). Since then, Tampa Bay on the Gulf Coast has lost more than 44 percent of its coastal wetlands acreage; Lake Worth on the Atlantic Coast has had an 87 percent decrease in mangrove acreage over the past 40 years and, on the Indian River outlet near Port St. Lucie, 86 percent of the mangrove areas have been lost to fisheries since the 1940's (Florida Department of Environmental Protection, 2017). The mangroves were removed for development of beach resorts, hotels, condos, and other structures to attract tourism. Today the remaining mangroves are protected. The mangrove ecosystem traps and cycles various organic materials, chemical elements, and nutrients, and provides attachment surfaces for marine organisms such as oysters, which filter and clean water. Mangrove forests protect the shoreline from erosion, provide storm buffers, protect habitats and biodiversity (Florida Department of Environmental Protection, 2017).

In Lemon Bay on the Gulf Coast of Florida, a seawall was built to protect residents from flooding, eliminating the red mangroves that used to attract marine life there (Meszaros, 2017). A new initiative, the brainchild of architect Keith van der Riet and biologist Jessene Aquino Thomas, is to create an artificial mangrove and attach it to the seawall. The artificial mangrove is engineered to attract oysters (to clean the water) and other marine life that used to live there. After six months, the team found small crabs, fish, and oyster larvae living on the artificial mangrove. The Weston WannaB Inn helped to fund the project in order to clean up the water, which had been experiencing red tides of algae blooms (Meszaros, 2017).

Another "natural" solution to beach erosion is dune restoration, which builds up coastal resilience. Broward County realized the value of dune restoration when it passed a law requiring builders of new developments and renovation projects along the coast to fund renourishment project using dunes. The South Florida dune restoration project is organized by the South Florida Audobon Society and the Youth Environmental Alliance (YEA) Project ROC (Reclaiming our Coastline). Lee Gottlieb of YEA gathers groups of volunteers on the weekends to plant sea oats in sand brought in to create dunes. The dunes are inconspicuous (no blocked ocean views) and the sea oats send down deep roots that anchor the sand and make the dune resistant to high winds and waves. Dunes are being used to protect beach access point in Miami Dade and Broward Counties on the Atlantic Coast (Gottlieb, 2017).

TECHNOLOGICAL INNOVATIONS IN EMERGENCY RESPONSE

Microgrids

Microgrids, a technology used by the military, has an innovative new role to play for civilians. The Center for Climate and Energy Solutions (C2ED, formerly known as the Pew Center for Global Climate Change) published an article and convened two national panels recently about the advantages of microgrids. C2ES (Vine et al., 2017, p. 1) defines a microgrid as

...a relatively small, controllable power system composed of one or more generation units connected to nearby load that can be operated with, or independently from, the local distribution and bulk (i.e. high-voltage) transmission system, referred to as the "macrogrid."

Microgrids keep the power on during outages, as they can run on renewables, natural gas-fueled combustion turbines, or emerging sources such as fuel cells or small modular nuclear reactors. They can be used as backup or as the main electricity source for a hospital, university, or neighborhood (Vine et al., 2017).

Drones

There is a new technology that has been shown to be effective in search and rescue missions and in firefighting—namely, the unmanned drone. From 2000 to 2013, almost 300 U.S. wildfire firefighters were killed while on duty (U.S. Fire Administration, 2013). The use of drones to inspect the safety of buildings during fires has helped to improve firefighter safety. In 2015, the Greater Manchester Fire and Rescue Service in the UK became the first force in the UK to use drone support around the clock. Their team of trained operators has permission to fly the unmanned aircraft 400 feet (122meters) above ground and 1,640 feet (500 meters) away from the point of drone control. The drones are equipped with infrared cameras to help firefighters fight nighttime blazes. The GMFRS has found the drones to be a great aid in improving the ground safety of firefighting crews; drones are being added to the emergency toolkit of fire rescue services around the UK (BBC, 2016).

Drones have also been found to be useful in protecting the safety of emergency response teams during search and recovery after extreme weather events. In the aftermath of earthquakes, hurricanes, floods and other natural hazards, rescue workers are exposed to physical and psychological hazards (Thacker et al., 2008; Noyes et al., 2009). For example, in April 2016, a magnitude-7.8 earthquake hit Ecuador, damaging homes, buildings and infrastructure and killing more than 660 people. Thousands of buildings and homes were damaged or destroyed; there was severe damage to infrastructure such as roadways and bridges. The earthquake caused the ground to shift, increasing the risk of landslides during 189 aftershocks in the two days after the quake (FSD, 2016). Drones were utilized by the government to collect data on how the ground had shifted and to assess the risk of landslides to help make decisions about how to proceed with rescue missions. The drones enabled emergency response personnel to see remote areas beyond roadways. The data collected were mapped into ortho-mosaics that were used to prioritize and target rescue missions (FSD, 2016). One caveat with drone use is the need to respect personal privacy; data collected should be encrypted for usage by authorized personnel.

GIS Platforms

An innovative technology that is helpful in all four phases of emergency management—preparation, mitigation, response, and recovery—is the

Geographic Information Systems (GIS) tool. GIS mapping helps to foster resilience by enabling communities to prepare ahead of time, operate effectively during events, and recover more quickly. The world leader in GIS systems is esri.com. Its ArcGIS platform provides contextual tools for mapping and spatial reasoning to identify locations and connect information (Esri.com, 2017).

NGOs are developing software to interface with municipal systems to create urban resilience. One such NGO is GEOcommand for a Safer City, which provides cloud-based software to first responders that use state-of-the-art GIS technologies from esri.com. All four phases of emergency management are covered. The multi-layer GIS platform provides situational awareness for fire, police, public works, local utility companies, and community planners. Geocommand's City Aware tool was deployed in a Massachusetts town whose fire departments (FDs) needed a daily technology tool that could be scaled up to handle major events and was secure enough to manage data sharing and prevent unauthorized access. The tool's GIS mapping enabled them to modernize their system with digitized, searchable, easily updateable, and quickly accessed plans (GEOcommand, 2016).

EDUCATION IS THE FOUNDATION OF A NEW WORLD ORDER

In the past several years we have witnessed record mean global temperatures and CO_2 concentrations not seen before in the industrial era; we have been experiencing changes in weather patterns, including extreme precipitation, droughts and wildfires, extreme storms and flooding, and desertification. The acidification of our oceans is threatening coral reefs and some species of marine life. We are moving toward an inundation of our coastlines that will cause tens of millions of climate refugees by the end of the century. The world is moving into uncharted territory. We need to learn to navigate a world that seemed unimaginable 50 years ago.

Education is key to making a successful transition to this new world order. Without it, the world could easily succumb to panic and warfare, as 100s of millions of climate refugees compete for scarce resources. Education is needed to make people understand climate change, to teach them how to conserve resources and use new technologies to adapt to its health impacts, and to convince them to change their energy consumption habits so as to mitigate its long-term effects.

Education brings people across the globe together; long-distance (online) education is already being used to train a 21st century healthcare workforce to serve in areas of critical shortages (Weiss, 2016). Rapidly spreading Internet access in formerly remote areas of Asia and Africa is facilitating the promotion of public health campaigns to help achieve the SDGs of eliminating poverty, hunger, and preventable premature deaths from lack of prenatal or other medical care. The Internet and maritime satellite telephones are technological tools that help responders locate vulnerable populations in need of humanitarian and medical aid.

A new world order requires a significant shift in how people consume natural resources. Such a shift requires a change in perceptions. The use of DOI theory (Rogers, 1962, 1983, 1995, 2003), described more fully in Chapter 4, helps to facilitate such a shift in mindset. It can be used in conjunction with the Transtheoretical Model Stages of Change (Prochaska et al., 1994, 2009), a psychological model of behavior change that helps to move people past denial to action. In the case of global warming, some deniers associate it with polar bears, not with their own lives. Others rationalize that it is "natural," not human-induced, and therefore they cannot do anything about it. The Stages of Change Model states that, to get someone past denial, you have to engage in consciousness raising—that is, providing education about the problem. Climate change education should begin in grade school and continue at the workplace and in the community. Peer role models can be very effective teachers One such role model making a big impact in south Florida is Delaney Reynolds, a high school senior who founded the Sink or Swim Project (Reynolds, 2017) to reach students in south Florida about the urgent need to reduce GHG emissions and slow down global warming. A passionate and dedicated advocate and speaker, Delaney has reached out through her website, books, and participation in "Saving Miami," the *Years of Living Dangerously* TV special with Jack Black for National Geographic (National Geographic, 2017).

CONCLUSION

Although the outlook is grim, we must not give up hope. Climate change cannot be stopped, but it can be slowed down, and new technologies can aid us in adapting to its impacts. There is power in numbers, as Bill McKibben of 350.org has frequently noted. Remember how the growing public groundswell of opinion against apartheid in South Africa eventually ended it? First came consciousness raising as the world was shown pictures of poverty and injustice under apartheid. Then the public had to be persuaded to act; mass demonstrations and media campaigns kept the pressure on South Africa to end apartheid. Finally, economic sanctions were implemented at universities and colleges, where students lobbied to force their institutions to divest all monies invested in South Africa. That action brought apartheid to an end. We need to put an end to climate deniers. We need to educate communities, from the youngest to the oldest, about what global warming is and how they can adapt to it with the aid of innovative technologies.

Economic incentives need to be adopted to promote the use of clean, renewable energy, sustainable agriculture and manufacturing, green building. Subsidies of fossil fuels and unsustainable farm practices must stop. Pollution controls must be enforced, with no back pumping of fertilizer runoff into our lakes! Industrial hazardous waste must be disposed of legally and safely. Carbon taxes and carbon capture and geologic sequestration should be used to reduce GHG emissions. Microgrids, drones, and GIS platforms can help us to build climate resilience

Sea level rise cannot be ignored any longer! Coastal communities need to be supported in their efforts to implement protective measures such as shoreline armoring and beach renourishment, and to elevate certain roadways and install pumps. Development must be limited in high-risk areas. We must acknowledge the reality of the long-term inevitable inundation of coastal areas, especially in the vulnerable and highly exposed state of Florida. Relocation plans need to be developed.

We all need to pull together to slow down global warming. To be resilient, we need to be socially cohesive and 100 percent unified in our efforts to reduce GHG emissions. We can do it! Our planet depends on it.

REFERENCES

100 Resilient Cities. (2017). Urban resilience. New York: Rockefeller Foundation. Retrieved from http://www.100resilientcities.org/resilience#/-_/

Allianz. (2017). Seven ways to reduce CO2 emissions. Munich, Germany. Retrieved from https://www.allianz.com/en/about_us/open-knowledge/topics/environment/articles/090226-seven-ways-to-reduce-co2-emissions.html/

Before the Flood. (2016). A film by Leonardo DiCaprio and Fisher Stevens. National Geographic and Ratpac Documentary Films. Retrieved from https://www.beforetheflood.com/

Bristol Resilience Strategy. (2016). Retrieved from http://lghttp.60358.nexcesscdn.net/8046264/images/page/-/100rc/pdfs/Bristol_Strategy_PDF.pdf

British Broadcasting Service (BBC). (2016). Manchester fire service uses drone to fight blazes (with video). Retrieved from http://www.bbc.com/news/34637892

Carbon Tax Center. (2017). What's a carbon tax? New York: Carbon Tax Center. Retrieved from https://www.carbontax.org/whats-a-carbon-tax/

Centers for Disease Control and Prevention (CDC). (2015). CDC's Building Resilience Against Climate Effects (BRACE) Framework. Atlanta, GA: CDC. Retrieved from https://www.cdc.gov/climateandhealth/brace.htm

Clayton, T. (2016). Changing before our eyes: natural consequences of sea level rise in Florida. In A. C. Hine, D. P. Chambers, T. D. Clayton, M. R. Hafen, & G. L. Mitchum (Eds.), *Sea Level Rise in Florida: Science, Impacts, and Options*. Gainesville, FL: University Press of Florida.

Climate & Health Meeting. (2017). Climate Reality Project. Retrieved from https://www.climaterealityproject.org/health#watch

Distel, O. (2016, June 27). Israel's water revolution and its overflow for the world. *Times of Israel Blog*. Retrieved from http://blogs.timesofisrael.com/israels-water-revolution-and-its-overflow-for-the-world/

Dobson, J. E., & Campbell, J. S. (2014). The flatness of U.S. states. *Geographical Review*, *104*(1), 1–9. doi:10.1111/j.1931-0846.2014.12001.x

Esri.com. (2017). Work smarter with ArcGIS, the mapping and analytic platform. Retrieved from http://www.esri.com/arcgis/about-arcgis

European Environment Agency (EEA). (2017). Global and European sea level. Copenhagen, Denmark: EEA. Retrieved from https://www.eea.europa.eu/data-and-maps/indicators/sea-level-rise-4/assessment-2

Florida Department of Environmental Protection (DEP). (2017). Florida's mangroves. Retrieved from https://softlive.dep.state.fl.us/fco/fco/content/floridas-mangroves

Florida Oceans and Coastal Council. (2010). *Climate Change and Sea Level Rise in Florida: Updates on the Effects of Climate Change on Florida's Ocean and Coastal Resources*. Retrieved from http://www.floridaoceanscouncil.org/reports/Climate_Change_and_Sea_Level_Rise.pdf

FSD. (2016). Using drones to map post-earthquake damage in Ecuador. Case study #13. Retrieved from http://drones.fsd.ch/en/case-study-no-13-using-drones-to-inspect-post-earthquake-damage-in-ecuador/

GEOcommand. (2016). Preparation, Mitigation, Response, Recovery. Retrieved from http://www.geocommand.com/

Global CSS. Institute. (2016). The Global Status of CCS: 2016 Summary Report.

Gore, A. (1992). *Earth in the Balance*. New York: Rodale.

Gore, A. (2006). *An Inconvenient Truth: The Planetary Emergency of Global Warming and What We Can Do About It. Documentary film and book*. New York: Rodale.

Gottlieb, L. (2017, January 17). Dune Restoration. Lecture presented in Plantation, FL, Davie, FL, Youth Environmental Alliance. Retrieved from http://www.yeafrog.org/community-restorations/

Grossman, M. M. (2016, June 28). Israel leading a "water revolution" in arid California. Jerusalem Post. Retrieved from http://www.jpost.com/Business-and-Innovation/Environment/Israel-leading-a-water-revolution-in-arid-California-457930

Hand, M. (2017). Bipartisan caucus members reject Trump agenda, introduce bill to create climate panel. Retrieved from https://thinkprogress.org/lawmakers-aim-to-create-climate-commission-43518aa563db

Hansen, J., Sato, M., Hearty, P., Ruedy, R., Kelley, M., Masson-Delmotte, V., & Lo, K.-W. et al. (2015). Ice melt, sea level rise and superstorms: Evidence from paleoclimate data, climate modeling, and modern observations that 2 degree global warming is highly dangerous. *Atmospheric Chemistry and Physics Discussion*, *15*(14), 20059–20179. doi:10.5194/acpd-15-20059-2015

International Energy Agency (IEA). (2013). Technology roadmap. Carbon capture and storage. Retrieved from http://www.iea.org/publications/freepublications/publication/TechnologyRoadmapCarbonCaptureandStorage.pdf

International Energy Agency (IEA). (2016). Global EV Outlook 2016: Beyond one million electric cars. Retrieved from http://www.iea.org/publications/freepublications/publication/Global_EV_Outlook_2016.pdf

International Energy Agency (IEA). (2017). IEA finds CO2 emissions flat for third straight year even as Global economy grew in 2016. Retrieved from https://www.iea.org/newsroom/news/2017/march/iea-finds-co2-emissions-flat-for-third-straight-year-even-as-global-economy-grew.html

IPCC. (2014). Summary for Policymakers. In O. Edenhofer et al. (Eds.), Climate Change 2014: Mitigation of Climate Change. Contribution of Working Group III to the Fifth Assessment Report of the Intergovernmental Panel on Climate Change. UK: Cambridge University Press.

JohnEnglander.net. (2017). Historic Trends of Global Temperature, CO_2 in the Atmosphere, and Sea Level. Retrieved from http://www.johnenglander.net/

Jolly, D. (2015, October 16). Norway Is a Model for Encouraging Electric Car Sales. The New York Times. Retrieved from https://www.nytimes.com/2015/10/17/business/international/norway-is-global-model-for-encouraging-sales-of-electric-cars.html?_r=0

Kershner, I. (2015, June 1). Israel's water revolution. *Mosaic Magazine*. Retrieved from https://mosaicmagazine.com/picks/2015/06/israels-water-revolution/

Klinenberg, E. (2002). Dying alone: an interview with Eric Klinenberg. Chicago: University of Chicago Press; Available at http://www.press.uchicago.edu/Misc/Chicago/443213in.html

Klinenberg, E. (2013). Adaptation: How can cities be "climate-proofed?" The New Yorker, January 7, 2013. New York: Conde Nast. Retrieved from http://www.newyorker.com/magazine/2013/01/07/adaptation-2

Lacis, A. A., Hansen, J. E., Russell, G. L., Oinas, V., & Jonas, J. (2013). The role of long-lived greenhouse gases as principal LW control knob that governs the global surface temperature for past and future climate change. *Tellus. Series B, Chemical and Physical Meteorology*, *65*(1), 19734. doi:10.3402/tellusb.v65i0.19734

Lacis, A. A., Schmidt, G. A., Rind, D., & Ruedy, R. A. (2010). Atmospheric CO2: Principal control15 knob governing Earths temperature. *Science*, *330*(6002), 356–359. doi:10.1126/science.1190653 PMID:20947761

Maibach, E. (2017, February 15). Panel: Communicating the Climate-Health Connection. Climate and Health Meeting. Climate Reality Project.

McKibben, B. (1989). *The End of Nature*. New York: Time Books.

McKibben, B. (2012, July 19). Global warming's terrifying new math. *Rolling Stone*. Retrieved from http://www.rollingstone.com/politics/news/global-warmings-terrifying-new-math-20120719

McKibben, B. (2013). Do the Math (the movie). New York: 350.org & PF Pictures.

McMichael, A. (2010). Urban Environments, Climate Change and Human Health: Risks, Adaptation, and Mitigation Co-Benefits. Canberra, ACT 0200 Australia: University of Australia. Retrieved from https://www.atse.org.au/Documents/International%20Colloboration/Workshops/Aust%20China%20Science%20and%20Tech/Climate%20Change/McMichael.pdf

Meszaros, J. (2017). Artificial Mangroves Could Bring Back Vanishing Habitats. In Florida. Miami, FL: WLRN. Retrieved from http://wlrn.org/post/artificial-mangroves-could-bring-back-vanishing-habitats-florida

Mirani, L. (2015). Norway's electric car incentives were so good they had to be stopped. Washington, D.C.: Quartz Media LLC. Retrieved from https://qz.com/400277/norway-electric-car-incentives-were-so-good-they-had-to-be-stopped/

National Geographic. (2017). Saving Miami with Jack Black. Years of Living Dangerously. National Geographic Channel. Retrieved from http://yearsoflivingdangerously.com/story/the-rising/

National Oceanic Atmospheric Administration (NOAA). (2015). What is LIDAR? Washington, D.C.: The Department of Commerce. Retrieved from http://oceanservice.noaa.gov/facts/lidar.html

National Oceanic Atmospheric Administration (NOAA). (2017). Trends in Atmospheric Carbon Dioxide. Washington, D.C.: The Department of Commerce. Retrieved from https://www.esrl.noaa.gov/gmd/ccgg/trends/

News, S. B. S. (2016, December 9). Factbox: Carbon taxes and emission trading schemes around the world. Retrieved from http://www.sbs.com.au/news/article/2016/12/08/factbox-carbon-taxes-and-emission-trading-schemes-around-world

Noyes, P.D., McElwee, M.K., Miller, H.D., Clark, B.W., Van Tiem, L.A., … & Levin, E.D. (2009). The toxicology of climate change: Environmental contaminants in a warming world. *Environment International, 35*(6), 971-986. doi:10.1016/j.envint.2009.02.006

Peeples, D. (2017). Miami's enhanced resilience plan: a lesson for other cities? Portland, OR: Smart Cities Council. Retrieved from http://smartcitiescouncil.com/article/miamis-enhanced-resilience-plan-lesson-other-cities

Prochaska, J. O., Norcross, J. C., & DiClemente, C. C. (1994). Changing for good. New York: Morrow.

Prochaska, J. O., Redding, C., & Evers, K. (2009). The Transtheoretical Model and stages of change. In K. Glanz, F. M. Lewis, & B. K. Rimer (Eds.), *Health Behavior and Health Education: Theory, Research and Practice* (4th ed.). Jossey-Bass Publications, Inc.

Reynolds, D. (2017). Sink or Swim Project. Retrieved from http://miamisearise.com/category/delaney- reynolds/

Rogers, E. M. (1962). *Diffusion of Innovations* (1st ed.). New York: Free Press.

Rogers, E. M. (1983). *Diffusion of Innovations* (3rd ed.). New York: Free Press.

Rogers, E. M. (1995). *Diffusion of Innovations* (4th ed.). New York: Free Press.

Rogers, E. M. (2003). *Diffusion of Innovations* (5th ed.). New York: Free Press.

South Florida Regional Planning Council. (2013). Adaptation Action Areas: Policy options for adaptive planning for rising sea levels. Southeast Florida Climate Compact. Retrieved from http://www.southeastfloridaclimatecompact.org/wp-content/uploads/2014/09/final-report-aaa.pdf

Stevens, F. (2016). *Before the Flood* (final press notes). Documentary film by Leonardo DiCaprio and Fisher Stevens, Martin Scorsese, Executive Producer. Ratpac Documentary Films. *National Geographic*.

Thacker, M.T.F., Lee, R., Sabogal, R.I., & Henderson, A. (2008). Overview of deaths associated with natural events, United States, 1979–2004. *Disasters, 32*, 303-315. doi:<ALIGNMENT.qj></ALIGNMENT>10.1111/j.1467-7717.2008.01041.x

UNESCO. (2013, October 21-23). Beijing Declaration on Building Learning Cities Lifelong Learning for All: Promoting Inclusion, Prosperity and Sustainability in Cities. *Proceedings of the International Conference on Learning Cities*, Beijing, China. UNESCO.

UNESCO. (2015). *UNESCO Global Network of Learning Cities. Guiding documents*. Hamburg, Germany: UNESCO Institute for Lifelong Learning.

United Nations. (2014a). United Nations Framework Convention on Climate Change. Retrieved from http://unfccc.int/essential_background/convention/items/6036.ph

United Nations. (2014b). *World urbanization prospects, 2014 revision*. New York: United Nations.

United Nations. (2015). Paris Agreement. Retrieved from http://unfccc.int/files/essential_background/convention/application/pdf/english_paris_agreement.pdf

United Nations. (2016). Marrakech Partnership for Global Action. New York: United Nations. Retrieved from http://unfccc.int/files/paris_agreement/application/pdf/marrakech_partnership_for_global_climate_action.pdf

United States Department of Energy (DOE). (2017). *Overview of carbon storage research*. Washington, D.C.: US DOE.

United States Environmental Protection Agency (EPA). (2016a). Carbon dioxide capture and sequestration. Retrieved from https://www3.epa.gov/climatechange/ccs/index.html

United States Environmental Protection Agency (EPA). (2016b). Inventory of U.S. Greenhouse Gas Emissions and Sinks, 1990-2014. Retrieved from https://www.epa.gov/ghgemissions/inventory-us-greenhouse-gas-emissions-and-sinks-1990-2014

United States Fire Administration (USFA), Federal Emergency Management Agency. (2013). Firefighter Fatalities in the United States in 2012. Retrieved from https://www.usfa.fema.gov/downloads/pdf/publications/ff_fat12.pdf

Vine, D., Attanasio, D., & Shittu, E. (2017). Microgrid Momentum: Building Efficient, Resilient Power. Arlington, VA: C2ES.

Wong, E. (2017, March 24). China plans a "green necklace" around Beijing to fight smog. *The New York Times*. New York: Arthur Ochs Sulzberger, Jr.

World Bank. (2015). State and Trends of Carbon Pricing. Washington, DC: World Bank Group/Ecofys. Retrieved from http://documents.worldbank.org/curated/en/636161467995665933/pdf/99533-REVISED- PUB-P153405-Box393205B.pdf

KEY TERMS AND DEFINITIONS

Adaptation: Adjustments in natural or human systems in response to actual or anticipated climatic stimuli or their impacts, with the goal of moderating harm and exploiting beneficial opportunities.

Carbon Capture and (Geologic) Sequestration: Capturing CO_2 from power plants or industrial processes, transporting it (usually in pipelines), and injecting it underground for storage into deep underground rock formations.

Carbon Credit: Any tradable certificate or permit representing the right to emit one ton of carbon dioxide or the mass of another greenhouse gas with a carbon dioxide equivalent (tCO_2e) equivalent to one ton of carbon dioxide.

Carbon Footprint: The amount of carbon dioxide (CO_2) a person produces in daily life.

Carbon Neutrality: Achieving a net zero carbon footprint by balancing a measured amount of carbon released with an equivalent amount sequestered or offset, or buying enough carbon credits to offset the difference.

Carbon Offsetting: The counteracting of carbon dioxide emissions with an equivalent reduction of carbon dioxide in the atmosphere.

Carbon Pricing: Charging users who emit carbon dioxide (CO_2) for their emissions, with the aim of reducing global CO_2 emissions.

Carbon Tax: A fee for making users of fossil fuels pay for climate damage their fuel use imposes by releasing carbon dioxide into the atmosphere, and for motivating a switch to clean energy.

Emission Trading System (ETS): A government-mandated, market-based approach to controlling pollution that provides economic incentives for reducing the emissions of pollutants.

Microgrid: Electricity distribution systems containing loads and distributed energy resources, (such as distributed generators, storage devices, or controllable loads) that can be operated in a controlled, coordinated way either while connected to the main power network or while islanded; they provide key resilience function in an outage.

Mitigation: Reducing and stabilizing the levels of heat-trapping greenhouse gases in the atmosphere.

Smog: A hazy atmospheric condition that occurs when ground-level ozone is created by chemical reactions between sunlight and oxides of nitrogen (NOx) and volatile organic compounds (VOC) emitted by vehicles, power plants, industrial boilers, refineries, chemical plants, and other sources.

Urban Resilience: The capacity of individuals, communities, institutions, businesses, and systems within a city to survive and grow no matter what kinds of chronic stresses and acute shocks they experience.

Related Readings

To continue IGI Global's long-standing tradition of advancing innovation through emerging research, please find below a compiled list of recommended IGI Global book chapters and journal articles in the areas of energy planning, climate change, and efficient energy. These related readings will provide additional information and guidance to further enrich your knowledge and assist you with your own research.

Adler, M. (2015). Floods Monitoring. In C. Maftei (Ed.), *Extreme Weather and Impacts of Climate Change on Water Resources in the Dobrogea Region* (pp. 312–344). Hershey, PA: IGI Global. doi:10.4018/978-1-4666-8438-6.ch011

Afzal, S. (2016). Implementation of Flooding Free Routing in Smart Grid: VCP Routing in Smart Gird. In A. Ahmad & N. Hassan (Eds.), *Smart Grid as a Solution for Renewable and Efficient Energy* (pp. 298–322). Hershey, PA: IGI Global. doi:10.4018/978-1-5225-0072-8.ch013

Ahmad, S., Ahmad, A., & Yaqub, R. (2016). Optimized Energy Consumption and Demand Side Management in Smart Grid. In A. Ahmad & N. Hassan (Eds.), *Smart Grid as a Solution for Renewable and Efficient Energy* (pp. 1–25). Hershey, PA: IGI Global. doi:10.4018/978-1-5225-0072-8.ch001

Arbaiy, N., Watada, J., & Lin, P. (2016). Fuzzy Random Regression-Based Modeling in Uncertain Environment. In P. Vasant & N. Voropai (Eds.), *Sustaining Power Resources through Energy Optimization and Engineering* (pp. 127–146). Hershey, PA: IGI Global. doi:10.4018/978-1-4666-9755-3.ch006

Arhin, A. (2016). Improving Sustainability of the Environment in a Changing Climate: Can REDD+ Rise to the Challenge? In S. Dinda (Ed.), *Handbook of Research on Climate Change Impact on Health and Environmental Sustainability* (pp. 327–346). Hershey, PA: IGI Global. doi:10.4018/978-1-4666-8814-8.ch016

Bahinipati, C. S., Patnaik, U., & Viswanathan, P. K. (2016). What Causes Economic Losses from Natural Disasters in India? In S. Dinda (Ed.), *Handbook of Research on Climate Change Impact on Health and Environmental Sustainability* (pp. 157–175). Hershey, PA: IGI Global. doi:10.4018/978-1-4666-8814-8.ch008

Banerjee, S. (2016). Social Innovation: A Theoretical Approach in Intertwining Climate Change with Social Innovation. In S. Dinda (Ed.), *Handbook of Research on Climate Change Impact on Health and Environmental Sustainability* (pp. 593–618). Hershey, PA: IGI Global. doi:10.4018/978-1-4666-8814-8.ch029

Barakabitze, A. A., Fue, K. G., Kitindi, E. J., & Sanga, C. A. (2016). Developing a Framework for Next Generation Integrated Agro Food-Advisory Systems in Developing Countries. *International Journal of Information Communication Technologies and Human Development*, *8*(4), 13–31. doi:10.4018/IJICTHD.2016100102

Basu, J. P. (2016). Coastal Poverty, Resource-Dependent Livelihood, Climate Change, and Adaptation: An Empirical Study in Indian Coastal Sunderbans. In S. Dinda (Ed.), *Handbook of Research on Climate Change Impact on Health and Environmental Sustainability* (pp. 441–454). Hershey, PA: IGI Global. doi:10.4018/978-1-4666-8814-8.ch022

Bekele, I., & Ganpat, W. (2015). Education, Extension, and Training for Climate Change. In W. Ganpat & W. Isaac (Eds.), *Impacts of Climate Change on Food Security in Small Island Developing States* (pp. 361–388). Hershey, PA: IGI Global. doi:10.4018/978-1-4666-6501-9.ch012

Bhaskar, A., Rao, G. B., & Vencatesan, J. (2017). Characterization and Management Concerns of Water Resources around Pallikaranai Marsh, South Chennai. In P. Rao & Y. Patil (Eds.), *Reconsidering the Impact of Climate Change on Global Water Supply, Use, and Management* (pp. 102–121). Hershey, PA: IGI Global. doi:10.4018/978-1-5225-1046-8.ch007

Bhatt, R. (2017). Zero Tillage for Mitigating Global Warming Consequences and Improving Livelihoods in South Asia. In W. Ganpat & W. Isaac (Eds.), *Environmental Sustainability and Climate Change Adaptation Strategies* (pp. 126–161). Hershey, PA: IGI Global. doi:10.4018/978-1-5225-1607-1.ch005

Bit, J., & Banerjee, S. (2016). Sustainable Forest Use and India's Economic Growth: A Structural Decomposition Analysis of Direct Forest Intensity. In S. Dinda (Ed.), *Handbook of Research on Climate Change Impact on Health and Environmental Sustainability* (pp. 306–326). Hershey, PA: IGI Global. doi:10.4018/978-1-4666-8814-8.ch015

Boonkerd, K. (2017). Development and Modification of Natural Rubber for Advanced Application. In T. Kobayashi (Ed.), *Applied Environmental Materials Science for Sustainability* (pp. 44–76). Hershey, PA: IGI Global. doi:10.4018/978-1-5225-1971-3.ch003

Boonmahitthisud, A. (2017). Natural Rubber and Rubber Blend Nanocomposites: Reinforcement of Natural Rubber with Polymer-Encapsulated Inorganic Nanohybrid Particles. In T. Kobayashi (Ed.), *Applied Environmental Materials Science for Sustainability* (pp. 77–105). Hershey, PA: IGI Global. doi:10.4018/978-1-5225-1971-3.ch004

Bostanci, S. H., & Albayrak, A. N. (2017). The Role of Eco-Municipalities in Climate Change for a Sustainable Future. In W. Ganpat & W. Isaac (Eds.), *Environmental Sustainability and Climate Change Adaptation Strategies* (pp. 213–231). Hershey, PA: IGI Global. doi:10.4018/978-1-5225-1607-1.ch008

Buta, C., Omer, I., & Andronic, A. (2015). Hydrological Risk Phenomena and Flood Analysis: Study Case – Taita Catchment, Romania. In C. Maftei (Ed.), *Extreme Weather and Impacts of Climate Change on Water Resources in the Dobrogea Region* (pp. 284–311). Hershey, PA: IGI Global. doi:10.4018/978-1-4666-8438-6.ch010

Carrillo, K. L., & Kobayashi, T. (2017). Natural Material Source of Bagasse Cellulose and Their Application to Hydrogel Films. In T. Kobayashi (Ed.), *Applied Environmental Materials Science for Sustainability* (pp. 19–43). Hershey, PA: IGI Global. doi:10.4018/978-1-5225-1971-3.ch002

Cazacu, G. B. (2015). Dobrogea Geology. In C. Maftei (Ed.), *Extreme Weather and Impacts of Climate Change on Water Resources in the Dobrogea Region* (pp. 73–118). Hershey, PA: IGI Global. doi:10.4018/978-1-4666-8438-6.ch004

Chai-Ittipornwong, T. (2017). Participation Framework to Sustainability: The Undercurrents in Bottled-Water Production and Consumption. In P. Rao & Y. Patil (Eds.), *Reconsidering the Impact of Climate Change on Global Water Supply, Use, and Management* (pp. 272–293). Hershey, PA: IGI Global. doi:10.4018/978-1-5225-1046-8.ch015

Chatterjee, T., & Dinda, S. (2016). Climate Change, Human Health and Some Economic Issues. In S. Dinda (Ed.), *Handbook of Research on Climate Change Impact on Health and Environmental Sustainability* (pp. 26–41). Hershey, PA: IGI Global. doi:10.4018/978-1-4666-8814-8.ch002

Chaudhary, N., & Pisolkar, Y. (2017). Issues, Concerns, and Local Stakes: Future of Water Resources in Coastal Villages of Devbag and Tarkarli, Coastal Maharashtra, India. In P. Rao & Y. Patil (Eds.), *Reconsidering the Impact of Climate Change on Global Water Supply, Use, and Management* (pp. 50–69). Hershey, PA: IGI Global. doi:10.4018/978-1-5225-1046-8.ch004

Chelcea, S., Ionita, M., & Adler, M. (2015). Identification of Dry Periods in the Dobrogea Region. In C. Maftei (Ed.), *Extreme Weather and Impacts of Climate Change on Water Resources in the Dobrogea Region* (pp. 52–72). Hershey, PA: IGI Global. doi:10.4018/978-1-4666-8438-6.ch003

Cohen, J. E., Clarke-Harris, D. O., Khan, A., & Isaac, W. P. (2015). Sustainable Management of Invasive Species for Small Island Developing States under Changing Climates. In W. Ganpat & W. Isaac (Eds.), *Impacts of Climate Change on Food Security in Small Island Developing States* (pp. 312–360). Hershey, PA: IGI Global. doi:10.4018/978-1-4666-6501-9.ch011

Das, S. (2016). Health Impact of Water-Related Diseases in Developing Countries on Account of Climate Change – A Systematic Review: A Study in Regard to South Asian Countries. In S. Dinda (Ed.), *Handbook of Research on Climate Change Impact on Health and Environmental Sustainability* (pp. 42–60). Hershey, PA: IGI Global. doi:10.4018/978-1-4666-8814-8.ch003

David, I., Beilicci, E., & Beilicci, R. (2015). Basics for Hydraulic Modelling of Flood Runoff Using Advanced Hydroinformatic Tools. In C. Maftei (Ed.), *Extreme Weather and Impacts of Climate Change on Water Resources in the Dobrogea Region* (pp. 205–239). Hershey, PA: IGI Global. doi:10.4018/978-1-4666-8438-6.ch008

Deenapanray, P. N., & Ramma, I. (2015). Adaptations to Climate Change and Climate Variability in the Agriculture Sector in Mauritius: Lessons from a Technical Needs Assessment. In W. Ganpat & W. Isaac (Eds.), *Impacts of Climate Change on Food Security in Small Island Developing States* (pp. 130–165). Hershey, PA: IGI Global. doi:10.4018/978-1-4666-6501-9.ch005

Dinda, S. (2016). Adaptation to Climate Change for Sustainable Development: A Survey. In S. Dinda (Ed.), *Handbook of Research on Climate Change Impact on Health and Environmental Sustainability* (pp. 363–391). Hershey, PA: IGI Global. doi:10.4018/978-1-4666-8814-8.ch018

Dinda, S. (2016). Climate Change, Trade Competitiveness, and Opportunity for Climate Friendly Goods in SAARC and Asia Pacific Regions. In S. Dinda (Ed.), *Handbook of Research on Climate Change Impact on Health and Environmental Sustainability* (pp. 515–536). Hershey, PA: IGI Global. doi:10.4018/978-1-4666-8814-8.ch026

Dongol, D., Bollin, E., & Feldmann, T. (2016). Battery Management Based on Predictive Control and Demand-Side Management: Smart Integration of Renewable Energy Sources. In A. Ahmad & N. Hassan (Eds.), *Smart Grid as a Solution for Renewable and Efficient Energy* (pp. 149–180). Hershey, PA: IGI Global. doi:10.4018/978-1-5225-0072-8.ch007

Drăguşin, D. (2015). Drought Effects on Groundwater in Dobrogea Plateau. In C. Maftei (Ed.), *Extreme Weather and Impacts of Climate Change on Water Resources in the Dobrogea Region* (pp. 119–144). Hershey, PA: IGI Global. doi:10.4018/978-1-4666-8438-6.ch005

Elsayed, A. M., Dakkama, H. J., Mahmoud, S., Al-Dadah, R., & Kaialy, W. (2017). Sustainable Cooling Research Using Activated Carbon Adsorbents and Their Environmental Impact. In T. Kobayashi (Ed.), *Applied Environmental Materials Science for Sustainability* (pp. 186–221). Hershey, PA: IGI Global. doi:10.4018/978-1-5225-1971-3.ch009

Emmanuel, M., Muasa, L., Chen, C., Mutisya, F., & Avtar, R. (2016). Impact of Rapid Urbanization and Climate Change on Agricultural Productivity in Africa: Climate Change Policies in the Agricultural Sector. In S. Dinda (Ed.), *Handbook of Research on Climate Change Impact on Health and Environmental Sustainability* (pp. 416–426). Hershey, PA: IGI Global. doi:10.4018/978-1-4666-8814-8.ch020

Emran, A., Rob, M. A., & Kabir, M. H. (2017). Coastline Change and Erosion-Accretion Evolution of the Sandwip Island, Bangladesh. *International Journal of Applied Geospatial Research*, *8*(2), 33–44. doi:10.4018/IJAGR.2017040103

Eudoxie, G., & Roopnarine, R. (2017). Climate Change Adaptation and Disaster Risk Management in the Caribbean. In W. Ganpat & W. Isaac (Eds.), *Environmental Sustainability and Climate Change Adaptation Strategies* (pp. 97–125). Hershey, PA: IGI Global. doi:10.4018/978-1-5225-1607-1.ch004

Eudoxie, G. D., & Wuddivira, M. (2015). Soil, Water, and Agricultural Adaptations. In W. Ganpat & W. Isaac (Eds.), *Impacts of Climate Change on Food Security in Small Island Developing States* (pp. 255–279). Hershey, PA: IGI Global. doi:10.4018/978-1-4666-6501-9.ch009

Fernández, F. J., Jiménez, A. D., Manzano, F. S., & Márquez, J. M. (2017). An Energy Management Strategy and Fuel Cell Configuration Proposal for a Hybrid Renewable System with Hydrogen Backup. *International Journal of Energy Optimization and Engineering*, *6*(1), 1–22. doi:10.4018/IJEOE.2017010101

Fernando, Y., & Wah, W. X. (2016). Moving forward a Parsimonious Model of Eco-Innovation: Results from a Content Analysis. In S. Dinda (Ed.), *Handbook of Research on Climate Change Impact on Health and Environmental Sustainability* (pp. 619–631). Hershey, PA: IGI Global. doi:10.4018/978-1-4666-8814-8.ch030

Garrick, T. A., & Liburd, O. E. (2017). Impact of Climate Change on a Key Agricultural Pest: Thrips. In W. Ganpat & W. Isaac (Eds.), *Environmental Sustainability and Climate Change Adaptation Strategies* (pp. 232–254). Hershey, PA: IGI Global. doi:10.4018/978-1-5225-1607-1.ch009

Goundar, S., & Appana, S. (2017). Mainstreaming Development Policies for Climate Change in Fiji: A Policy Gap Analysis and the Role of ICTs. In W. Ganpat & W. Isaac (Eds.), *Environmental Sustainability and Climate Change Adaptation Strategies* (pp. 1–31). Hershey, PA: IGI Global. doi:10.4018/978-1-5225-1607-1.ch001

Gupta, A. C. (2016). Bioeconomic Fishery Management: Changing Paradigms towards Eco-System Based Management. In S. Dinda (Ed.), *Handbook of Research on Climate Change Impact on Health and Environmental Sustainability* (pp. 261–281). Hershey, PA: IGI Global. doi:10.4018/978-1-4666-8814-8.ch013

Hiremath, R., Kumar, B., Bansode, S. S., Nulkar, G., Patil, S. S., & Murali, J. (2017). Industrial Wastewater Management in the Context of Climate Change Adaptation in Selected Cities of India: A Business Approach. In P. Rao & Y. Patil (Eds.), *Reconsidering the Impact of Climate Change on Global Water Supply, Use, and Management* (pp. 294–313). Hershey, PA: IGI Global. doi:10.4018/978-1-5225-1046-8.ch016

Huyen, P. T. (2017). Clay Minerals Converted to Porous Materials and Their Application: Challenge and Perspective. In T. Kobayashi (Ed.), *Applied Environmental Materials Science for Sustainability* (pp. 141–164). Hershey, PA: IGI Global. doi:10.4018/978-1-5225-1971-3.ch007

Iese, V., Maeke, J., Holland, E., Wairiu, M., & Naidu, S. (2015). Farming Adaptations to the Impacts of Climate Change and Extreme Events in Pacific Island Countries: Case Study of Bellona Atoll, Solomon Islands. In W. Ganpat & W. Isaac (Eds.), *Impacts of Climate Change on Food Security in Small Island Developing States* (pp. 166–194). Hershey, PA: IGI Global. doi:10.4018/978-1-4666-6501-9.ch006

Ikematsu, S., Tada, I., & Nagasaki, Y. (2017). Analysis of Lipids Produced by Microalgae Isolated from the Area around Okinawa, Japan. In T. Kobayashi (Ed.), *Applied Environmental Materials Science for Sustainability* (pp. 222–233). Hershey, PA: IGI Global. doi:10.4018/978-1-5225-1971-3.ch010

Inogwabini, B. (2017). Congo Basin's Shrinking Watersheds: Potential Consequences on Local Communities. In P. Rao & Y. Patil (Eds.), *Reconsidering the Impact of Climate Change on Global Water Supply, Use, and Management* (pp. 211–226). Hershey, PA: IGI Global. doi:10.4018/978-1-5225-1046-8.ch012

Ionita, M., & Chelcea, S. (2015). Spatio-Temporal Variability of Seasonal Drought over the Dobrogea Region. In C. Maftei (Ed.), *Extreme Weather and Impacts of Climate Change on Water Resources in the Dobrogea Region* (pp. 17–51). Hershey, PA: IGI Global. doi:10.4018/978-1-4666-8438-6.ch002

Issakhov, A. (2016). Mathematical Modelling of the Thermal Process in the Aquatic Environment with Considering the Hydrometeorological Condition at the Reservoir-Cooler by Using Parallel Technologies. In P. Vasant & N. Voropai (Eds.), *Sustaining Power Resources through Energy Optimization and Engineering* (pp. 227–243). Hershey, PA: IGI Global. doi:10.4018/978-1-4666-9755-3.ch010

Javid, T. (2016). Geographic Information System for the Smart Grid. In A. Ahmad & N. Hassan (Eds.), *Smart Grid as a Solution for Renewable and Efficient Energy* (pp. 344–362). Hershey, PA: IGI Global. doi:10.4018/978-1-5225-0072-8.ch015

Juma, D. W., Reuben, M., Wang, H., & Li, F. (2017). Adaptive Coevolution: Realigning the Water Governance Regime to the Changing Climate. In P. Rao & Y. Patil (Eds.), *Reconsidering the Impact of Climate Change on Global Water Supply, Use, and Management* (pp. 314–325). Hershey, PA: IGI Global. doi:10.4018/978-1-5225-1046-8.ch017

Kais, S. M. (2017). Climate Change: Vulnerability and Resilience in Commercial Shrimp Aquaculture in Bangladesh. In W. Ganpat & W. Isaac (Eds.), *Environmental Sustainability and Climate Change Adaptation Strategies* (pp. 162–187). Hershey, PA: IGI Global. doi:10.4018/978-1-5225-1607-1.ch006

Kamboj, V. K., & Bath, S. K. (2016). Scope of Biogeography-Based Optimization for Economic Load Dispatch and Multi-Objective Unit Commitment Problem. In P. Vasant & N. Voropai (Eds.), *Sustaining Power Resources through Energy Optimization and Engineering* (pp. 360–389). Hershey, PA: IGI Global. doi:10.4018/978-1-4666-9755-3.ch015

Karmaoui, A. (2016). Environmental Vulnerability to Climate Change in Mediterranean Basin: Socio-Ecological Interactions between North and South. In S. Dinda (Ed.), *Handbook of Research on Climate Change Impact on Health and Environmental Sustainability* (pp. 105–138). Hershey, PA: IGI Global. doi:10.4018/978-1-4666-8814-8.ch006

Khai, H. V. (2016). Assessing Urban Residents' Willingness to Pay for Preserving the Biodiversity of Swamp Forest. In S. Dinda (Ed.), *Handbook of Research on Climate Change Impact on Health and Environmental Sustainability* (pp. 283–305). Hershey, PA: IGI Global. doi:10.4018/978-1-4666-8814-8.ch014

Khalid, S. (2016). Application of Adaptive Tabu Search Algorithm in Hybrid Power Filter and Shunt Active Power Filters: Application of ATS Algorithm in HPF and APF. In P. Vasant & N. Voropai (Eds.), *Sustaining Power Resources through Energy Optimization and Engineering* (pp. 276–308). Hershey, PA: IGI Global. doi:10.4018/978-1-4666-9755-3.ch012

Khan, S. S. (2016). Modeling and Operating Strategies of Micro-Grids for Renewable Energy Communities. In A. Ahmad & N. Hassan (Eds.), *Smart Grid as a Solution for Renewable and Efficient Energy* (pp. 97–122). Hershey, PA: IGI Global. doi:10.4018/978-1-5225-0072-8.ch005

Khanna, B. K. (2017). Indian National Strategy for Climate Change Adaptation and Mitigation. In W. Ganpat & W. Isaac (Eds.), *Environmental Sustainability and Climate Change Adaptation Strategies* (pp. 32–63). Hershey, PA: IGI Global. doi:10.4018/978-1-5225-1607-1.ch002

Khanna, B. K. (2017). Vulnerability of the Lakshadweep Coral Islands in India and Strategies for Mitigating Climate Change Impacts. In W. Ganpat & W. Isaac (Eds.), *Environmental Sustainability and Climate Change Adaptation Strategies* (pp. 64–96). Hershey, PA: IGI Global. doi:10.4018/978-1-5225-1607-1.ch003

Kobayashi, T. (2017). Introduction of Environmental Materials. In T. Kobayashi (Ed.), *Applied Environmental Materials Science for Sustainability* (pp. 1–18). Hershey, PA: IGI Global. doi:10.4018/978-1-5225-1971-3.ch001

Kumar, C. P. (2016). Impact of Climate Change on Groundwater Resources. In S. Dinda (Ed.), *Handbook of Research on Climate Change Impact on Health and Environmental Sustainability* (pp. 196–221). Hershey, PA: IGI Global. doi:10.4018/978-1-4666-8814-8.ch010

Kumar, K. V., Kumar, S. S., Selvakumar, A. I., & Kumar, R. S. (2016). Recent Techniques to Identify the Stator Fault Diagnosis in Three Phase Induction Motor. In P. Vasant & N. Voropai (Eds.), *Sustaining Power Resources through Energy Optimization and Engineering* (pp. 309–325). Hershey, PA: IGI Global. doi:10.4018/978-1-4666-9755-3.ch013

Kumar, M. (2016). Environmental Sustainability in the Fashion Supply Chain in India. *International Journal of Social Ecology and Sustainable Development*, *7*(3), 1–33. doi:10.4018/IJSESD.2016070101

Kumar, M. (2017). Economic Evaluation of Solar Cooling Schemes. *International Journal of Energy Optimization and Engineering*, *6*(1), 23–48. doi:10.4018/IJEOE.2017010102

Kumar, R., Rao, P., & Areendran, G. (2017). Understanding Glacial Retreat in the Indian Himalaya: Historical Trends and Field Studies from a Large Glacier. In P. Rao & Y. Patil (Eds.), *Reconsidering the Impact of Climate Change on Global Water Supply, Use, and Management* (pp. 33–49). Hershey, PA: IGI Global. doi:10.4018/978-1-5225-1046-8.ch003

Kumari, S., & Patil, Y. (2017). Achieving Climate Smart Agriculture with a Sustainable Use of Water: A Conceptual Framework for Sustaining the Use of Water for Agriculture in the Era of Climate Change. In P. Rao & Y. Patil (Eds.), *Reconsidering the Impact of Climate Change on Global Water Supply, Use, and Management* (pp. 122–143). Hershey, PA: IGI Global. doi:10.4018/978-1-5225-1046-8.ch008

Kumari, S., Patil, Y., & Rao, P. (2017). An Approach to Sustainable Watershed Management: Case Studies on Enhancing Sustainability with Challenges of Water in Western Maharashtra. In P. Rao & Y. Patil (Eds.), *Reconsidering the Impact of Climate Change on Global Water Supply, Use, and Management* (pp. 252–271). Hershey, PA: IGI Global. doi:10.4018/978-1-5225-1046-8.ch014

Lallo, C. H., Smalling, S., Facey, A., & Hughes, M. (2017). The Impact of Climate Change on Small Ruminant Performance in Caribbean Communities. In W. Ganpat & W. Isaac (Eds.), *Environmental Sustainability and Climate Change Adaptation Strategies* (pp. 296–321). Hershey, PA: IGI Global. doi:10.4018/978-1-5225-1607-1.ch011

Lawrence, J., Simpson, L., & Piggott, A. (2015). Protected Agriculture: A Climate Change Adaptation for Food and Nutrition Security. In W. Ganpat & W. Isaac (Eds.), *Impacts of Climate Change on Food Security in Small Island Developing States* (pp. 196–220). Hershey, PA: IGI Global. doi:10.4018/978-1-4666-6501-9.ch007

Li, K., & Kobayashi, T. (2017). Ionic Liquids and Poly (Ionic Liquid)s Used as Green Solvent and Ultrasound Responded Materials. In T. Kobayashi (Ed.), *Applied Environmental Materials Science for Sustainability* (pp. 327–346). Hershey, PA: IGI Global. doi:10.4018/978-1-5225-1971-3.ch015

Liapakis, A., Costopoulou, C., Tsiligiridis, T., & Sideridis, A. (2017). Studying Corporate Social Responsibility Activities in the Agri-Food Sector: The Greek Case. *International Journal of Agricultural and Environmental Information Systems*, *8*(1), 1–13. doi:10.4018/IJAEIS.2017010101

Loi, N. K., Huyen, N. T., Tu, L. H., Tram, V. N., Liem, N. D., Dat, N. L., & Minh, D. N. et al. (2017). Sustainable Land Use and Watershed Management in Response to Climate Change Impacts: Case Study in Srepok Watershed, Central Highland of Vietnam. In W. Ganpat & W. Isaac (Eds.), *Environmental Sustainability and Climate Change Adaptation Strategies* (pp. 255–295). Hershey, PA: IGI Global. doi:10.4018/978-1-5225-1607-1.ch010

Londhe, S. (2016). Impact of Climate Change on Agriculture and Food Security. *International Journal of Disease Control and Containment for Sustainability*, *1*(1), 32–46. doi:10.4018/IJDCCS.2016010103

Londhe, S. (2017). Inter Linkages of Water, Climate, and Agriculture. In P. Rao & Y. Patil (Eds.), *Reconsidering the Impact of Climate Change on Global Water Supply, Use, and Management* (pp. 166–194). Hershey, PA: IGI Global. doi:10.4018/978-1-5225-1046-8.ch010

Londhe, S. L. (2016). Climate Change and Agriculture: Impacts, Adoption, and Mitigation. In S. Dinda (Ed.), *Handbook of Research on Climate Change Impact on Health and Environmental Sustainability* (pp. 393–415). Hershey, PA: IGI Global. doi:10.4018/978-1-4666-8814-8.ch019

Maftei, C., & Papatheodorou, K. (2015). Mathematical Models Used for Hydrological Floodplain Modeling. In C. Maftei (Ed.), *Extreme Weather and Impacts of Climate Change on Water Resources in the Dobrogea Region* (pp. 240–283). Hershey, PA: IGI Global. doi:10.4018/978-1-4666-8438-6.ch009

Maharaj, R., Singh-Ackbarali, D., & Sankat, C. K. (2015). Postharvest Management Strategies. In W. Ganpat & W. Isaac (Eds.), *Impacts of Climate Change on Food Security in Small Island Developing States* (pp. 221–254). Hershey, PA: IGI Global. doi:10.4018/978-1-4666-6501-9.ch008

Mahboob, Q. (2016). Identification of Reliability Critical Items in Large and Complex Rail Electrical Networks. In A. Ahmad & N. Hassan (Eds.), *Smart Grid as a Solution for Renewable and Efficient Energy* (pp. 226–248). Hershey, PA: IGI Global. doi:10.4018/978-1-5225-0072-8.ch010

Mahmood, W. A., & Azarian, M. H. (2017). Inorganic-Organic Composite Materials from Liquid Natural Rubber and Epoxidised Natural Rubber Derivatives: Prospects and Applications. In T. Kobayashi (Ed.), *Applied Environmental Materials Science for Sustainability* (pp. 128–140). Hershey, PA: IGI Global. doi:10.4018/978-1-5225-1971-3.ch006

Majumder, J., Shah, P., & Kumar, S. (2016). Heat Stress Vulnerability among Indian Workmen. In S. Dinda (Ed.), *Handbook of Research on Climate Change Impact on Health and Environmental Sustainability* (pp. 61–80). Hershey, PA: IGI Global. doi:10.4018/978-1-4666-8814-8.ch004

Maximay, S. (2015). The Caribbean's Response to Climate Change Impacts. In W. Ganpat & W. Isaac (Eds.), *Impacts of Climate Change on Food Security in Small Island Developing States* (pp. 33–66). Hershey, PA: IGI Global. doi:10.4018/978-1-4666-6501-9.ch002

Meldrum, H. M., Szymanski, D., Oches, E. A., & Davis, P. T. (2016). A Picture is Worth a Thousand Words: Commentary of Broadcast Meteorologists on the Visual Presentation of Climate Change. *International Journal of Social Ecology and Sustainable Development*, *7*(4), 1–16. doi:10.4018/IJSESD.2016100101

Mili, B., Barua, A., & Katyaini, S. (2016). Climate Change and Adaptation through the Lens of Capability Approach: A Case Study from Darjeeling, Eastern Himalaya. In S. Dinda (Ed.), *Handbook of Research on Climate Change Impact on Health and Environmental Sustainability* (pp. 455–469). Hershey, PA: IGI Global. doi:10.4018/978-1-4666-8814-8.ch023

Minea, G., Bandoc, G., & Neculau, G. (2015). Seasonal Statistical Variability of Precipitations in Dobrogea and Danube Delta. In C. Maftei (Ed.), *Extreme Weather and Impacts of Climate Change on Water Resources in the Dobrogea Region* (pp. 1–16). Hershey, PA: IGI Global. doi:10.4018/978-1-4666-8438-6.ch001

Minhas, D. M., & Hussain, S. (2016). Efficient Control Strategies to Optimize Electricity Cost and Consumer Satisfaction. In A. Ahmad & N. Hassan (Eds.), *Smart Grid as a Solution for Renewable and Efficient Energy* (pp. 69–96). Hershey, PA: IGI Global. doi:10.4018/978-1-5225-0072-8.ch004

Mohapatra, M., & Baladhandautham, C. B. (2016). Implementation of Improved Control Strategy of DC-AC Converter using Delta-Sigma Modulator. In A. Ahmad & N. Hassan (Eds.), *Smart Grid as a Solution for Renewable and Efficient Energy* (pp. 249–272). Hershey, PA: IGI Global. doi:10.4018/978-1-5225-0072-8.ch011

Moustache, A. M. (2015). Adaptation to Impacts of Climate Change on the Food and Nutrition Security Status of a Small Island Developing State: The Case of the Republic of Seychelles. In W. Ganpat & W. Isaac (Eds.), *Impacts of Climate Change on Food Security in Small Island Developing States* (pp. 96–129). Hershey, PA: IGI Global. doi:10.4018/978-1-4666-6501-9.ch004

Mujere, N., & Moyce, W. (2017). Climate Change Impacts on Surface Water Quality. In W. Ganpat & W. Isaac (Eds.), *Environmental Sustainability and Climate Change Adaptation Strategies* (pp. 322–340). Hershey, PA: IGI Global. doi:10.4018/978-1-5225-1607-1.ch012

Mukherjee, S., & Chakraborty, D. (2016). Does Fiscal Policy Influence Per Capita CO2 Emission?: A Cross Country Empirical Analysis. In S. Dinda (Ed.), *Handbook of Research on Climate Change Impact on Health and Environmental Sustainability* (pp. 568–592). Hershey, PA: IGI Global. doi:10.4018/978-1-4666-8814-8.ch028

Murtaza, G., Saqib, M., Saifullah, Rehman, M. Z., Naveed, M., & Ghafoor, A. (2017). Mitigation of Climate Change Impacts through Treatment and Management of Low Quality Water for Irrigation in Pakistan. In P. Rao, & Y. Patil (Eds.), Reconsidering the Impact of Climate Change on Global Water Supply, Use, and Management (pp. 84-101). Hershey, PA: IGI Global. doi:10.4018/978-1-5225-1046-8.ch006

N'Yeurt, A. D., & Iese, V. (2015). Marine Plants as a Sustainable Source of Agri-Fertilizers for Small Island Developing States (SIDS). In W. Ganpat & W. Isaac (Eds.), *Impacts of Climate Change on Food Security in Small Island Developing States* (pp. 280–311). Hershey, PA: IGI Global. doi:10.4018/978-1-4666-6501-9.ch010

Nersesian, R. L., & Strang, K. D. (2017). Feasibility Approaches to Reduce the Unreliability of Gas, Nuclear, Coal, Solar and Wind Electricity Production. *International Journal of Risk and Contingency Management*, *6*(1), 54–69. doi:10.4018/IJRCM.2017010104

Ngubevana, L. (2017). Sustainable Development Dilemmas of Biofuels Research and Production: A Snapshot in South Africa. *International Journal of Energy Optimization and Engineering*, *6*(2), 24–41. doi:10.4018/IJEOE.2017040102

Nyangon, J., Alabbas, N., & Agbemabiese, L. (2017). Entangled Systems at the Energy-Water-Food Nexus: Challenges and Opportunities. In P. Rao & Y. Patil (Eds.), *Reconsidering the Impact of Climate Change on Global Water Supply, Use, and Management* (pp. 144–165). Hershey, PA: IGI Global. doi:10.4018/978-1-5225-1046-8.ch009

Onutai, S., Jiemsirilers, S., & Kobayashi, T. (2017). Geopolymer Sourced with Fly Ash and Industrial Aluminum Waste for Sustainable Materials. In T. Kobayashi (Ed.), *Applied Environmental Materials Science for Sustainability* (pp. 165–185). Hershey, PA: IGI Global. doi:10.4018/978-1-5225-1971-3.ch008

Osmani, A. R. (2016). Greenhouse Gas Mitigation through Energy Efficiency: Perform, Achieve, and Trade (PAT) – India's Emission Trading Scheme. In S. Dinda (Ed.), *Handbook of Research on Climate Change Impact on Health and Environmental Sustainability* (pp. 537–566). Hershey, PA: IGI Global. doi:10.4018/978-1-4666-8814-8.ch027

Osmani, A. R. (2017). Tipaimukh Multipurpose Hydroelectric Project: A Policy Perspective – Indo-Bangla Priorities, Indigenous Peoples' Rights, and Environmental Concerns. In P. Rao & Y. Patil (Eds.), *Reconsidering the Impact of Climate Change on Global Water Supply, Use, and Management* (pp. 227–251). Hershey, PA: IGI Global. doi:10.4018/978-1-5225-1046-8.ch013

Ozpinar, A., & Ozil, E. (2016). Smart Grid and Demand Side Management: Application of Metaheuristic and Artificial Intelligence Algorithms. In A. Ahmad & N. Hassan (Eds.), *Smart Grid as a Solution for Renewable and Efficient Energy* (pp. 49–68). Hershey, PA: IGI Global. doi:10.4018/978-1-5225-0072-8.ch003

Padigala, B. S. (2017). Traditional Water Management System for Climate Change Adaptation in Mountain Ecosystems. In P. Rao & Y. Patil (Eds.), *Reconsidering the Impact of Climate Change on Global Water Supply, Use, and Management* (pp. 9–32). Hershey, PA: IGI Global. doi:10.4018/978-1-5225-1046-8.ch002

Page, T. (2017). A Feasibility Study in Energy Harvesting from Piezoelectric Keyboards. *International Journal of Energy Optimization and Engineering*, *6*(2), 1–23. doi:10.4018/IJEOE.2017040101

Paul, S., & Roy, P. K. (2016). A Novel Optimization Algorithm for Transient Stability Constrained Optimal Power Flow. In P. Vasant & N. Voropai (Eds.), *Sustaining Power Resources through Energy Optimization and Engineering* (pp. 147–176). Hershey, PA: IGI Global. doi:10.4018/978-1-4666-9755-3.ch007

Pieroni, A., & Iazeolla, G. (2016). Engineering QoS and Energy Saving in the Delivery of ICT Services. In P. Vasant & N. Voropai (Eds.), *Sustaining Power Resources through Energy Optimization and Engineering* (pp. 208–226). Hershey, PA: IGI Global. doi:10.4018/978-1-4666-9755-3.ch009

Polprasert, J., Ongsakul, W., & Dieu, V. N. (2016). Improved Pseudo-Gradient Search Particle Swarm Optimization for Optimal Power Flow Problem. In P. Vasant & N. Voropai (Eds.), *Sustaining Power Resources through Energy Optimization and Engineering* (pp. 177–207). Hershey, PA: IGI Global. doi:10.4018/978-1-4666-9755-3.ch008

Potiyaraj, P. (2017). Poly (Lactic Acid) Generated for Advanced Materials. In T. Kobayashi (Ed.), *Applied Environmental Materials Science for Sustainability* (pp. 106–127). Hershey, PA: IGI Global. doi:10.4018/978-1-5225-1971-3.ch005

Rale, V., & Tendulkar, P. (2017). Common Duckweeds as a Model System for Climate Change Impact Assessment. In P. Rao & Y. Patil (Eds.), *Reconsidering the Impact of Climate Change on Global Water Supply, Use, and Management* (pp. 364–372). Hershey, PA: IGI Global. doi:10.4018/978-1-5225-1046-8.ch019

Ramakrishna, A., & Bang, S. (2015). The Impacts of Climate Change on Food Security and Management in Papua New Guinea. In W. Ganpat & W. Isaac (Eds.), *Impacts of Climate Change on Food Security in Small Island Developing States* (pp. 67–95). Hershey, PA: IGI Global. doi:10.4018/978-1-4666-6501-9.ch003

Rani, S. (2016). Assessment of Annual, Monthly, and Seasonal Trends in the Long Term Rainfall of the Garhwal Himalayas. In S. Dinda (Ed.), *Handbook of Research on Climate Change Impact on Health and Environmental Sustainability* (pp. 222–241). Hershey, PA: IGI Global. doi:10.4018/978-1-4666-8814-8.ch011

Rao, P., & Patil, Y. (2017). Recent Trends, Issues, and Challenges in Water Resource Development and Global Climate Change. In P. Rao & Y. Patil (Eds.), *Reconsidering the Impact of Climate Change on Global Water Supply, Use, and Management* (pp. 1–8). Hershey, PA: IGI Global. doi:10.4018/978-1-5225-1046-8.ch001

Ray, S. (2016). A Framework for Understanding Adaptation by Manufacturing Industries. In S. Dinda (Ed.), *Handbook of Research on Climate Change Impact on Health and Environmental Sustainability* (pp. 471–481). Hershey, PA: IGI Global. doi:10.4018/978-1-4666-8814-8.ch024

Roberts, T. G., & Rodriguez, M. T. (2015). An Overview of Climate Change and Impacts on Food Security in Small Island Developing States. In W. Ganpat & W. Isaac (Eds.), *Impacts of Climate Change on Food Security in Small Island Developing States* (pp. 1–31). Hershey, PA: IGI Global. doi:10.4018/978-1-4666-6501-9.ch001

Roşu, L., & Macarov, L. I. (2015). Management of Drought and Floods in the Dobrogea Region. In C. Maftei (Ed.), *Extreme Weather and Impacts of Climate Change on Water Resources in the Dobrogea Region* (pp. 403–443). Hershey, PA: IGI Global. doi:10.4018/978-1-4666-8438-6.ch013

Roşu, L., & Zăgan, R. (2015). Management of Drought and Floods in Romania. In C. Maftei (Ed.), *Extreme Weather and Impacts of Climate Change on Water Resources in the Dobrogea Region* (pp. 345–402). Hershey, PA: IGI Global. doi:10.4018/978-1-4666-8438-6.ch012

Roy, J., Ghosh, D., Mukhopadhyay, K., & Ghosh, A. (2016). Exacerbating Health Risks in India due to Climate Change: Rethinking Approach to Health Service Provision. In S. Dinda (Ed.), *Handbook of Research on Climate Change Impact on Health and Environmental Sustainability* (pp. 1–25). Hershey, PA: IGI Global. doi:10.4018/978-1-4666-8814-8.ch001

Roy, P. K. (2016). A Novel Evolutionary Optimization Technique for Solving Optimal Reactive Power Dispatch Problems. In P. Vasant & N. Voropai (Eds.), *Sustaining Power Resources through Energy Optimization and Engineering* (pp. 244–275). Hershey, PA: IGI Global. doi:10.4018/978-1-4666-9755-3.ch011

Roy, P. K., Dutta, S., & Nandi, D. (2016). Optimal Reactive Power Dispatch Incorporating TCSC-TCPS Devices Using Different Evolutionary Optimization Techniques. In P. Vasant & N. Voropai (Eds.), *Sustaining Power Resources through Energy Optimization and Engineering* (pp. 326–359). Hershey, PA: IGI Global. doi:10.4018/978-1-4666-9755-3.ch014

Roy, P. K., & Ghosh, M. (2017). Combined Heat and Power Dispatch using Hybrid Genetic Algorithm and Biogeography-based Optimization. *International Journal of Energy Optimization and Engineering*, *6*(1), 49–65. doi:10.4018/IJEOE.2017010103

Samanta, D. (2016). Lack of Land Tenure Security as Challenges to Sustainable Development: An Assessment in the Context of Bihar, India. In S. Dinda (Ed.), *Handbook of Research on Climate Change Impact on Health and Environmental Sustainability* (pp. 348–362). Hershey, PA: IGI Global. doi:10.4018/978-1-4666-8814-8.ch017

Sasirekha, S., & Swamynathan, S. (2017). Fuzzy Rule Based Environment Monitoring System for Weather Controlled Laboratories using Arduino. *International Journal of Intelligent Information Technologies*, *13*(1), 50–66. doi:10.4018/IJIIT.2017010103

Satoh, M. (2017). Metal Ion Separation with Functional Adsorbents and Phytoremediation Used as Sustainable Technologies. In T. Kobayashi (Ed.), *Applied Environmental Materials Science for Sustainability* (pp. 284–312). Hershey, PA: IGI Global. doi:10.4018/978-1-5225-1971-3.ch013

Scorza, F. (2016). Towards Self Energy-Management and Sustainable Citizens Engagement in Local Energy Efficiency Agenda. *International Journal of Agricultural and Environmental Information Systems*, *7*(1), 44–53. doi:10.4018/IJAEIS.2016010103

Sen, S. K., & Pookayaporn, J. (2017). Role of Water-Energy-Waste Inter-Relatedness to Drive Sustainability amid Climate Concerns. In P. Rao & Y. Patil (Eds.), *Reconsidering the Impact of Climate Change on Global Water Supply, Use, and Management* (pp. 195–210). Hershey, PA: IGI Global. doi:10.4018/978-1-5225-1046-8.ch011

Senapati, S., & Gupta, V. (2016). Impacts of Climate Change on Fish Productivity: A Quantitative Measurement. In S. Dinda (Ed.), *Handbook of Research on Climate Change Impact on Health and Environmental Sustainability* (pp. 243–260). Hershey, PA: IGI Global. doi:10.4018/978-1-4666-8814-8.ch012

Senapati, S., & Gupta, V. (2016). Vulnerability to Climate Change: Issues and Challenges towards Developing Vulnerability Indicator. In S. Dinda (Ed.), *Handbook of Research on Climate Change Impact on Health and Environmental Sustainability* (pp. 82–104). Hershey, PA: IGI Global. doi:10.4018/978-1-4666-8814-8.ch005

Serban, C., & Maftei, C. (2015). Using Grid Computing and Satellite Remote Sensing in Evapotranspiration Estimation. In C. Maftei (Ed.), *Extreme Weather and Impacts of Climate Change on Water Resources in the Dobrogea Region* (pp. 145–173). Hershey, PA: IGI Global. doi:10.4018/978-1-4666-8438-6.ch006

Shanmuganathan, S., Narayanan, A., & Medagoda, N. P. (2016). Temporal Data Analysis and Mining Methods for Modelling the Climate Change Effects on Malaysia's Oil Palm Yield at Different Regional Scales. In S. Dinda (Ed.), *Handbook of Research on Climate Change Impact on Health and Environmental Sustainability* (pp. 482–513). Hershey, PA: IGI Global. doi:10.4018/978-1-4666-8814-8.ch025

Singh, R. K. (2017). Impact of Climate Change on the Retreat of Himalayan Glaciers and Its Impact on Major River Hydrology: Himalayan Glacier Hydrology. In P. Rao & Y. Patil (Eds.), *Reconsidering the Impact of Climate Change on Global Water Supply, Use, and Management* (pp. 70–83). Hershey, PA: IGI Global. doi:10.4018/978-1-5225-1046-8.ch005

Singh-Ackbarali, D., & Maharaj, R. (2017). Mini Livestock Ranching: Solution to Reducing the Carbon Footprint and Negative Environmental Impacts of Agriculture. In W. Ganpat & W. Isaac (Eds.), *Environmental Sustainability and Climate Change Adaptation Strategies* (pp. 188–212). Hershey, PA: IGI Global. doi:10.4018/978-1-5225-1607-1.ch007

Siwe, A. T., & Tembine, H. (2016). Energy Cost Saving Tips in Distributed Power Networks. In A. Ahmad & N. Hassan (Eds.), *Smart Grid as a Solution for Renewable and Efficient Energy* (pp. 26–48). Hershey, PA: IGI Global. doi:10.4018/978-1-5225-0072-8.ch002

Srivastava, N. (2016). Climate Change Mitigation: Collective Efforts and Responsibly. In S. Dinda (Ed.), *Handbook of Research on Climate Change Impact on Health and Environmental Sustainability* (pp. 427–439). Hershey, PA: IGI Global. doi:10.4018/978-1-4666-8814-8.ch021

Stennikov, V., Barakhtenko, E., Sokolov, D., & Oshchepkova, T. (2016). Problems of Modeling and Optimization of Heat Supply Systems: New Methods and Software for Optimization of Heat Supply System Parameters. In P. Vasant & N. Voropai (Eds.), *Sustaining Power Resources through Energy Optimization and Engineering* (pp. 76–101). Hershey, PA: IGI Global. doi:10.4018/978-1-4666-9755-3.ch004

Stennikov, V., Oshchepkova, T., & Stennikov, N. (2016). Problems of Modeling and Optimization of Heat Supply Systems: Methods to Comprehensively Solve the Problem of Heat Supply System Expansion and Reconstruction. In P. Vasant & N. Voropai (Eds.), *Sustaining Power Resources through Energy Optimization and Engineering* (pp. 26–53). Hershey, PA: IGI Global. doi:10.4018/978-1-4666-9755-3.ch002

Stennikov, V., Penkovskii, A., & Khamisov, O. (2016). Problems of Modeling and Optimization of Heat Supply Systems: Bi-Level Optimization of the Competitive Heat Energy Market. In P. Vasant & N. Voropai (Eds.), *Sustaining Power Resources through Energy Optimization and Engineering* (pp. 54–75). Hershey, PA: IGI Global. doi:10.4018/978-1-4666-9755-3.ch003

Stennikov, V. A., & Postnikov, I. V. (2016). Problems of Modeling and Optimization of Heat Supply Systems: Methodological Support for a Comprehensive Analysis of Fuel and Heat Supply Reliability. In P. Vasant & N. Voropai (Eds.), *Sustaining Power Resources through Energy Optimization and Engineering* (pp. 102–126). Hershey, PA: IGI Global. doi:10.4018/978-1-4666-9755-3.ch005

Stone, R. J. (2017). Modelling the Frequency of Tropical Cyclones in the Lower Caribbean Region. In W. Ganpat & W. Isaac (Eds.), *Environmental Sustainability and Climate Change Adaptation Strategies* (pp. 341–349). Hershey, PA: IGI Global. doi:10.4018/978-1-5225-1607-1.ch013

Swain, M. (2016). Vulnerability to Local Climate Change: Farmers' Perceptions on Trends in Western Odisha, India. In S. Dinda (Ed.), *Handbook of Research on Climate Change Impact on Health and Environmental Sustainability* (pp. 139–155). Hershey, PA: IGI Global. doi:10.4018/978-1-4666-8814-8.ch007

Swain, M., & Swain, M. (2016). Evolution and Efficacy of Drought Management Policies and Programmes: The Case of Western Odisha, India. In S. Dinda (Ed.), *Handbook of Research on Climate Change Impact on Health and Environmental Sustainability* (pp. 176–194). Hershey, PA: IGI Global. doi:10.4018/978-1-4666-8814-8.ch009

Takahashi, Y. (2017). Eco-Friendly On-Site Water Analyses for Ultra-Trace Harmful Ions. In T. Kobayashi (Ed.), *Applied Environmental Materials Science for Sustainability* (pp. 313–326). Hershey, PA: IGI Global. doi:10.4018/978-1-5225-1971-3.ch014

Taokaew, S., Phisalaphong, M., & Newby, B. Z. (2017). Bacterial Cellulose: Biosyntheses, Modifications, and Applications. In T. Kobayashi (Ed.), *Applied Environmental Materials Science for Sustainability* (pp. 255–283). Hershey, PA: IGI Global. doi:10.4018/978-1-5225-1971-3.ch012

Tauch, S., Liu, W., & Pears, R. (2016). Measuring Cascading Failures in Smart Grid Networks. In A. Ahmad & N. Hassan (Eds.), *Smart Grid as a Solution for Renewable and Efficient Energy* (pp. 208–225). Hershey, PA: IGI Global. doi:10.4018/978-1-5225-0072-8.ch009

Trang, T. T. (2017). Study on the Use of Biomass Polymer Sheets in Water/Alcohol Pervaporation as a Sustainable Source of Alcohol Energy. In T. Kobayashi (Ed.), *Applied Environmental Materials Science for Sustainability* (pp. 234–254). Hershey, PA: IGI Global. doi:10.4018/978-1-5225-1971-3.ch011

Tsiaras, S. (2017). Exploring the Impact of Tourism to the Sustainable Development of Mountain Regions: Implications of the Climatic Conditions. *International Journal of Agricultural and Environmental Information Systems*, *8*(1), 14–28. doi:10.4018/IJAEIS.2017010102

Tugjamba, N., Yembuu, B., Gantumur, A., & Getsel, U. (2016). Policy Provisions and Teachers Needs on Climate Change Education for Sustainable Development in Mongolia. *International Journal of Asian Business and Information Management*, *7*(4), 36–48. doi:10.4018/IJABIM.2016100103

Tyukhov, I., Rezk, H., & Vasant, P. (2016). Modern Optimization Algorithms and Applications in Solar Photovoltaic Engineering. In P. Vasant & N. Voropai (Eds.), *Sustaining Power Resources through Energy Optimization and Engineering* (pp. 390–445). Hershey, PA: IGI Global. doi:10.4018/978-1-4666-9755-3.ch016

Uddin, Z., Shah, N., Ahmad, A., Mehmood, W., & Alam, F. (2016). Signal Processing Techniques in Smart Grids. In A. Ahmad & N. Hassan (Eds.), *Smart Grid as a Solution for Renewable and Efficient Energy* (pp. 273–297). Hershey, PA: IGI Global. doi:10.4018/978-1-5225-0072-8.ch012

Ventrapragada, E. A., & Rayavarapu, N. (2017). Climate Change and Agriculture: Time for a Responsive and Responsible System of Water Management. In P. Rao & Y. Patil (Eds.), *Reconsidering the Impact of Climate Change on Global Water Supply, Use, and Management* (pp. 326–363). Hershey, PA: IGI Global. doi:10.4018/978-1-5225-1046-8.ch018

Wang, Q., & Liu, H. (2017). Optimized Base Station Sleeping and Renewable Energy Procurement Scheme Using PSO. *International Journal of Swarm Intelligence Research*, *8*(1), 54–73. doi:10.4018/IJSIR.2017010103

Wong, J., & Lim, Y. S. (2016). Revolution of Energy Storage System in Smart Grids. In A. Ahmad & N. Hassan (Eds.), *Smart Grid as a Solution for Renewable and Efficient Energy* (pp. 181–206). Hershey, PA: IGI Global. doi:10.4018/978-1-5225-0072-8.ch008

Wu, Y., Tan, X., Qian, L. P., Tsang, D. H., Song, W., & Yu, L. (2016). Management of Scheduling and Trading in Hybrid Energy Trading Market. In A. Ahmad & N. Hassan (Eds.), *Smart Grid as a Solution for Renewable and Efficient Energy* (pp. 123–148). Hershey, PA: IGI Global. doi:10.4018/978-1-5225-0072-8.ch006

Xiang, M., Liu, W., Bai, Q., & Al-Anbuky, A. (2016). Dynamic Trust Elective Geo Routing to Secure Smart Grid Communication Networks. In A. Ahmad & N. Hassan (Eds.), *Smart Grid as a Solution for Renewable and Efficient Energy* (pp. 323–343). Hershey, PA: IGI Global. doi:10.4018/978-1-5225-0072-8.ch014

Zăgan, S., & Chiţu, M. (2015). The Influence of Air Temperature on the Quality Parameters of the Black Sea Coastal Waters. In C. Maftei (Ed.), *Extreme Weather and Impacts of Climate Change on Water Resources in the Dobrogea Region* (pp. 174–204). Hershey, PA: IGI Global. doi:10.4018/978-1-4666-8438-6.ch007

Zharkov, S. (2016). Assessment and Enhancement of the Energy Supply System Efficiency with Emphasis on the Cogeneration and Renewable as Main Directions for Fuel Saving. In P. Vasant & N. Voropai (Eds.), *Sustaining Power Resources through Energy Optimization and Engineering* (pp. 1–25). Hershey, PA: IGI Global. doi:10.4018/978-1-4666-9755-3.ch001

About the Author

Debra N. Weiss-Randall, EdD, MA, CHES, is a Guest Lecturer at Florida Atlantic University and a Consultant at Florida International University; she previously taught at the City University of New York and SUNY Purchase for 16 years. She is certified to design and teach online courses. Dr. Weiss-Randall holds a Masters' degree (2007) and Doctorate (2010) in Health Education from Teachers' College, Columbia University, and is a Certified Health Education Consultant through NCHEC. Her doctoral research, a statewide study of asthma management on New York college campuses, was replicated in Texas in 2014 and won a national award. In 2016, Dr. Weiss-Randall served as Project Manager for an environmental justice grant from the U.S. Environmental Protection Agency. Two vulnerable coastal communities were surveyed about perceived risk and impacts from sea level rise. Dr. Weiss-Randall is active in the Southeast Florida Regional Climate Compact, Sea Level Rise Solutions Project of South Florida, and South Florida Asthma Consortium. She also is the artistic director of the Debra Weiss Dance Company (DWDC), a not-for-profit organization that is known for its arts-in-education programs, including Fish Tales, which teaches children about the Hudson River estuary, and The Trees Are Singing, an earth day program about preserving our forests. In 2017, Dr. Weiss-Randall received a grant to start afterschool programs for at-risk youth in Palm Beach County, Florida.

Index

K

L

M

N

P

R

S

T

U

V

W

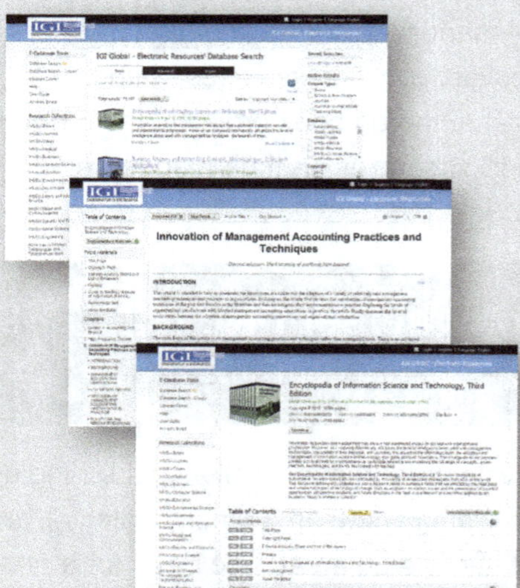

www.ingramcontent.com/pod-product-compliance
Lightning Source LLC
LaVergne TN
LVHW081525300926
849161LV00006B/1298